DISCARD

CHEMICAL TRANSMISSION
OF NERVE IMPULSES

CHEMICAL TRANSMISSION OF NERVE IMPULSES

A HISTORICAL SKETCH

Z. M. BACQ

PERGAMON PRESS

OXFORD · NEW YORK · TORONTO
SYDNEY · BRAUNSCHWEIG

Pergamon Press Offices:

U.K.	Pergamon Press Ltd., Headington Hill Hall, Oxford OX3 0BW, England
U.S.A.	Pergamon Press Inc., Maxwell House, Fairview Park, Elmsford, New York 10523, U.S.A.
CANADA	Pergamon of Canada, Ltd., 207 Queen's Quay West, Toronto 1, Canada
AUSTRALIA	Pergamon Press (Austr.) Pty. Ltd., 19a Boundary Street, Rushcutters Bay, N.S.W. 2011, Australia
FRANCE	Pergamon Press SARL, 24 rue des Ecoles, 75240 Paris, Cedex 05, France
WEST GERMANY	Pergamon Press GMbH, 3300 Braunschweig, Postfach 2923, Burgplatz 1, West Germany

Library of Congress Catalog Card No. 75-27296

Translated from the French book
'Les transmissions chimiques de l'influx nerveux'
published by Bordas Editeur © 1974

First English edition
Copyright © 1975 Pergamon Press Ltd.

Printed in Belgium by G. Thone

0 08 020512 7

'A lie may be written, a writing may be lost, but what the eye has seen is truth and remains in the mind.'

J. Conrad (The lagoon)

'Passons passons puisque tout passe Je me retournerai souvent.'

G. Apollinaire
(Cors de chasse)

Pour Fernande
Quarante années de patience.

Contents

Preface to the English edition IX

Preface to the French edition XI

1. Definition and importance of the phenomenon of chemical transmission 1

2. General organization of the nervous system in Vertebrates 6

3. The forerunners 11

4. Cholinergic transmission 15

   4.1. *First phase*. The parasympathetic division of the autonomic nervous system 15
      4.1.1. The fundamental experiments of Loewi on the Amphibian heart 15
      4.1.2. Extension of Loewi's concept to the whole parasympathetic innervation in Mammals 22
      4.1.3. Problems of nomenclature 26

   4.2. *Second phase*. The cholinergic nature of the motor nerves to striated muscles and the preganglionic fibres of the autonomic nervous system 27
      4.2.1. Striated muscle 31
      4.2.2 Preganglionic fibres 34

5. Adrenergic transmission 36

6. The progress of biochemical research   51

   6.1. Acetylcholine synthesis and studies of cholineste-
      rases   51

   6.2. Catecholamine synthesis   54

7. The opposition to the theory of chemical transmission   57

8. Some blind alleys   71

9. Comparative physiology   76

10. The triumph: The central nervous system   83

11. Working in Sir Henry Dale's department   88

12. Scientists and science   92

   12.1. Europe versus the U.S.A.   92

   12.2. The adventurous spirit; the flair   94

   12.3. The power and simplicity of facts and their
      utilization   96

   12.4. The diversity of temperaments   99

A few complementary references 105

Preface to the English edition

This little book is a translation of the French edition 'Les Transmissions chimiques de l'influx nerveux' published by Gauthier-Villars in Paris (1974) in a collection 'Discours de la Méthode', the Director of which is Boris Rybak.

The author and his collaborators regret that they were obliged to prepare this translation in a great hurry. Fortunately the printer (Imprimerie Thone in Liège, Belgium) has done a magnificent job and the final text may become a useful though very personal work.

Special thanks are also due to Robert Maxwell for having given the necessary instructions to get this edition ready in time for the celebration of Sir Henry Dale centenary.

Preface to the French edition

The concept of chemical transmission of nerve impulses by a few molecules endowed with special properties gained general acceptance by biologists during the years 1945-1950, and its spectacular development in the last two decades has introduced truly revolutionary ideas not only in physiology, but also in pharmacology and biophysics.

The students and young scientists of 1975 cannot imagine the difficulties that the promotion of this theory had to surmount between the years 1920 and 1950, before it was accepted. Thus, it may be of interest that a survivor of this crucial epoch took pains to tell the history of this concept, at least of its beginnings, since he was lucky enough to work for a long time with two of its main pioneers (Sir Henry Dale and W. B. Cannon), to have known Otto Loewi and to have been (or still be) the friend of numerous scientists who have devoted most of their time to accumulating experimental evidence in favour of the theory and to develop the magnificent avenues that it opened to research.

The other survivor who might contribute at first hand to this historical sketch is W. Feldberg; I thank him for the interest he has taken in my book, and for his useful information.

In a review article of a book by J. Needham, Sir Hans Krebs writes:

'The rapid growth and the ever increasing complexity of science makes the task of the historian a very formidable one. So I think that it is becoming increasingly important that those who have personally participated in the progress of science, or witnessed it from close quarters, should be encouraged to write histories of their own special field, as Keilin did in his history of Cell Respiration and Cytochrome. Others are unlikely to master adequately the intimate details.' (*Nature, 227,* 1061, 1970).

The author hopes that his text will be useful mainly because it pays tribute to the work, character and virtues of a small number of exceptional men, who have left their mark in this beautiful story.

The author cannot escape a feeling of dissatisfaction, even of frustration ; there are so many other anecdotes and other interesting events that he would have liked to report, but which could not be inserted within the wise but strict limits of this series of books.

The reader will find opinions expressed here which probably will give rise to some dispute. One cannot escape one's own temperament. Adventurers have definite features. Their errors are more revealing than their successes; one should be kind to them. I fear that the following text is not only incomplete, but also biased.

1

Definition and importance of the phenomenon of chemical transmission

Whether one considers a crocus bulb, a tree, a worm or a camel, the living organism is a distinct entity not only because of its obvious physical form, but also because of its functional integration. The biochemistry and biophysics of the millions of cells in a creature higher up the evolutionary scale are controlled (1) by a circulatory system, and (2) by a specialized system, the nervous system of Metazoa, which made its first appearance in Coelenterates (fresh water hydra, medusa, sea anemona) and acquired an extraordinary degree of development in Vertebrates and in certain Invertebrates (in the Cephalopods, for example).

For a long time biologists were preoccupied with the anatomical aspects of evolution. However, in recent years tremendous progress has been made in the field of biochemical evolution through molecular biology ; step by step we are able to follow the changes in the composition and structure of certain proteins, such as the cytochromes or the respiratory pigments of blood. There has also been an evolution in the physiological processes which through the use of a few very simple molecules ensure the transmission of nerve impulses.

The physiologists spent far too long in contrasting two systems of integration: (1) The hormones, well defined chemical substances which regulate the growth of plants and animals, the moulting of insects, the passage of ions and of water through cell membranes, etc., and (2) The neurones, the nerve cells which are peculiar to Metazoa. They receive messages which they pass on, modify or stop, and which enable the organism to react rapidly, often within a fraction of a second, to external stimuli or to changes taking place internally.

In the early decades of this century the physiologists liked to emphasize the differences between hormonal regulation, which is essentially slow and affects the whole organism, and neural regulation which is much faster and more precise, and is often confined to a small group of cells. To contrast hormonal and nervous control, however, is of no real value, except for teaching purposes. True, the endocrine glands, the hormone suppliers (thyroid, parathyroid, pancreas, ovaries, testis and adrenal cortex), have their own physiological functions and their regulatory mechanisms are up to a point relatively independent. But it is easy to demonstrate that the nervous system can interfere with these functions. On the other hand, the hormones can influence the function of the nervous system itself, as well as the reactions of muscles and glands to nerve stimulation.

The nervous system looks today like a highly sophisticated machine which—thanks to a series of ingenious mechanisms—can utilize for new purposes certain simple substances that make their appearance very early on in evolution. The concept of a neuro-endocrine system, as W. B. Cannon defined it in the 1920s, is still valid. The most striking example is the series of catecholamines (see p. 36), of which some (adrenaline, noradrenaline) are real hormones secreted into the blood by the adrenal medulla during emotional shock, physical exercise, cold, or when the blood sugar concentration drops below the point tolerable to nerve cells. These same hormones are also the chemical mediators or transmitters, the indispensible substances for the transmission of excitation in one class of nerve fibres which innervate the heart, smooth muscles and glands (see p. 43).

In order to grasp the importance of the phenomenon of chemical transmission in the nervous system, it is necessary to remind ourselves briefly of the physiology and anatomy of neurones, the main characteristics of which were defined by the famous Spanish histologist, Ramon y Cajal, at the end of the last century. We can give the following schematic description (see Fig. 1). As in all other cells, the neurone has a nucleus, a nucleolus, mitochondria, microsomes, lysosomes, and all the other usual intracellular structures. It has numerous relatively short processes called dendrites, which at specialized structural sites (spines, etc.) receive the impulses coming from the sense organs or from other nerve cells. It has, in addition, a single long process (sometimes very long—several metres in large mammals), the axon, which sends out the impulse.

Nerves are bundles of axons, which one can compare to telephone cables with their numerous individual wires. The nerve impulse is a wave of depolarisation which spreads rapidly (1-40 metres/sec.) along the nerve fibres. Its passage can be recorded by means of electrodes and suitable electronic amplification.

This wave stops at the end of the axon. The axon does not enter muscle cells or nerve cells ; there is no continuity between the axon and the cell that it innervates ; it makes contact with the membrane of the innervated cell through synaptic structures which have been well defined by electron microscopy and histochemistry. If one cuts the axon, the peripheral end now cut off from the body of the nerve cell degenerates. Usually the cells innervated by the axon do not atrophy and, at first sight do not differ in appearance from normally innervated cells, except that no nerve fibres are present. Mammals can live quite well with a totally denervated heart, which no longer has any connection with the central nervous system. But the mammalian skeletal muscle atrophies after denervation, and its structure changes for the following reasons : it becomes paralyzed and no longer receives certain trophic substances, which are different from the chemical transmitters and which control its structure and the properties of its excitable membranes.

A major problem which has occupied neurophysiologists for the last century is to understand how an impulse which arrives at the endings of an axon succeeds in affecting the cell membrane of muscle, glands or nerve cells. The two structures are separated by a definite space of 200 to 500 Å. The impulse must cross this

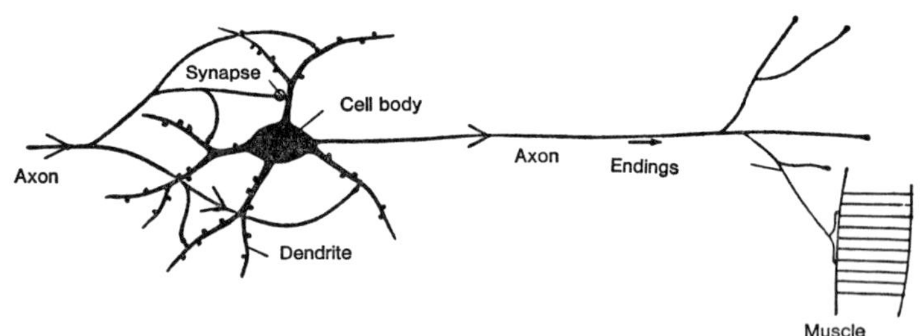

FIG. 1. — Schematic presentation of a neurone receiving endings from another neurone. The arrows indicate direction of the nerve impulses.

gap. In modern terms, the region where the nerve ending touches the cell is called the synapse; a distinction is made between the structures and the events which are *presynaptic* and which depend on the neurone, and those which are *postsynaptic* and are controlled by the effector cell.

As explained in the third chapter, until 1930-35 physiologists believed that it was the difference in the electric potential itself, the eddy current, which passed from the endings of the axon to the membrane of the effector cell and brought about the transmission. Today we know that with rare exceptions transmission of the nerve impulse is effected in all tissues (smooth muscles, glands, striated muscles, heart and nerve cells) through molecules which are stored in the presynaptic nerve endings. The electrical events are naturally linked to the release of these transmitters and we therefore speak in a general way of 'electro-chemical' events that occur at synapses. The neurone can, therefore, be looked upon as a kind of electrical device which transmits quanta of very active molecules from its axon terminals whenever a nerve impulse reaches them.

The intention of this modest piece of work is to show how this concept was arrived at. However, before getting to the heart of the subject matter, it may be useful to show those readers who are not experts how the nervous system is organized in Vertebrates.

<h1 style="text-align:center">2</h1>

General organization of the nervous system in Vertebrates

Schematically, Vertebrates can be said to be made up of a framework of bone, a group of soft tissues (muscles, glands, skin and connective tissue) and a neuro-endocrine system which controls the functioning of these tissues. The blood and lymph allows oxygen, nutrient materials and waste products to circulate without difficulty. Without the neuro-endocrine system a Vertebrate would not be able to act as a functional unit and a man would not be able to develop his unique personality.

In this publication, we shall consider only that part of the nervous system which is called the 'efferent' one, which consists of nerve fibres that leave the central nervous system (brain and spinal cord) and the ganglia of the autonomic nervous system, small collections of cells which in the embryo migrate from the central nervous system (CNS) and become anatomically separated from this system. For the control of the neuro-endocrine system, three parts are always required.

1. The sensory organs which consist not only of the sense organs for sight, hearing, smell, taste and touch, but also of sensory endings in the heart, blood vessels, muscles and tendons (they respond to stretch) and inform the central nervous system about the state of the organism and its environment. Certain chemosensitive cells in the carotid bodies respond to variations in the partial pressure of oxygen and carbon dioxide (CO_2) in

the blood. From the sensory organs nerve impulses pass from the periphery into the central nervous system. The impulses travel in a 'centripetal' or 'afferent' direction.

2. The nerve centres, well protected in the skull and the vertebral column receive these impulses, deal with them and then send them through well defined nerve tracks to ensure the integration.

3. The fibres which leave the centres are called efferent or motor; their action potentials, the impulses, travel in a 'centrifugal' direction. Those fibres which originate from the cells of the anterior horn of the spinal cord pass directly, without interruption, to the skeletal muscles which are under voluntary control. Other efferent fibres originating from the spinal cord are called preganglionic. They innervate the heart, smooth muscles (blood vessels, gastro-intestinal tract, uterus, bronchi, iris, etc.) as well as the glands (the salivary, gastric, intestinal, tear, sweat glands, etc.). Unlike the skeletal muscles these organs are not under voluntary control. We cannot prevent the heart from accelerating during an emotional shock, nor can we stop sweating when the weather is hot. This is the reason why these nerves are called 'autonomic'. All of our functions are under the strict control of reflexes, often very complicated ones, which are integrated in the CNS and ultimately modify the rate at which impulses travel along the pre- and postganglionic fibres.

Anatomically, the autonomic nervous system is divided into an *(ortho)sympathetic* and a *parasympathetic* system (Fig. 2). The sympathetic preganglionic fibres leave the spinal cord from the first thoracic to the second or third lumbar vertebra. They pass through a continuous chain of sympathetic ganglia (one ganglion for each vertebra) which are connected by nerve fibres. There are two chains of ganglia, that run under the pleura or the peritoneum, one on each side of the vertebral column; for this reason these ganglia are called paravertebral. Fibres of the sympathetic system extend beyond the upper and lower limits of the paravertebral chains and innervate the head and genital organs. They thus innervate every organ of the body. There is no vessel, muscle or gland that does not receive a sympathetic innervation.

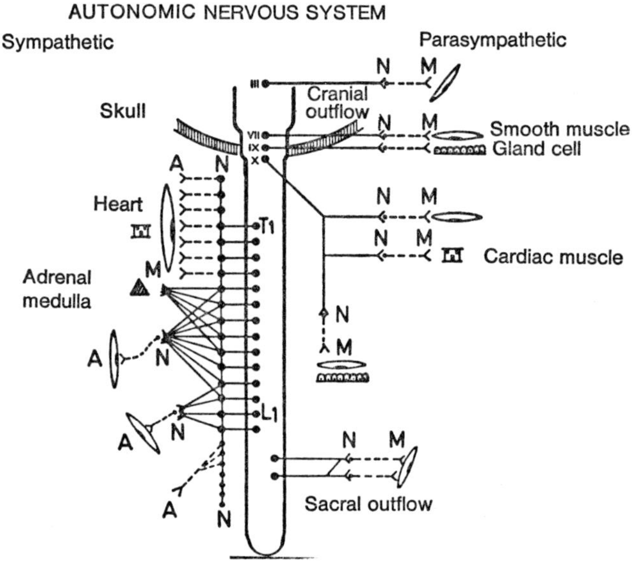

FIG. 2. — Diagram to indicate the organization of the autonomic nervous system. For simplicity the sympathetic division is shown on the right and the parasympathetic division on the left although innervation by both divisions is bilateral. Preganglionic fibres are shown in solid, postganglionic fibres in interrupted lines. All preganglionic fibres are cholinergic; they end in the ganglia where the nicotinic receptors (N) are located. The parasympathetic postganglionic fibres are cholinergic; they innervate heart, smooth muscles and glands; the receptors are muscarinic (M). The sympathetic postganglionic fibres are mostly adrenergic (A) with a few exceptions which are cholinergic (M), for example, those to the sweat glands of the cat. (From Graham, *Pharmacology for Medical Students*, Oxford University Press, 1966.)

The parasympathetic system is divided into two parts:

(1) The cranial division which consists of fibres running in the IIIrd, VIIth, IXth and Xth cranial nerves. The fibres in the IIIrd nerve go to the iris, those in the VIIth and IXth nerves innervate, among other organs, the salivary glands and the vessels of the tongue, and those in the Xth, the vagi, innervate heart, lungs, stomach and small intestine.

(2) The sacral division which leaves the sacral segments of the spinal cord and innervates the colon, genital organs and bladder.

As a rule the ganglion cells in the parasympathetic system, in contrast to the sympathetic ones, are located very close to the innervated cells, even within the organ. It is difficult, if not impossible, to place electrodes on the postganglionic fibres for electrical stimulation ; consequently we almost always deal with the preganglionic fibres. Many organs receive fibres from both the sympathetic and parasympathetic systems, the effects of which are generally antagonistic. This anatomical arrangement makes for an extremely delicate balance, and failure in one system may be compensated by the activity of the other.

Here are some examples. First the pupil; its diameter is controlled by two muscles of the iris: the *sphincter,* composed of circular fibres whose contraction produces a reduction in the size of the pupil and the *dilator* whose fibres dilate the pupil when they contract. The ocular parasympathetic nerve stimulates the constrictor, the sympathetic (whose fibres arise from the first thoracic segments and go up in the neck until they reach the important superior cervical ganglion) stimulates the dilator, while at the same time inhibiting the sphincter. Reflex dilatation of the pupil in response to a decrease in light is essentially due to a decrease in parasympathetic tone, i.e. to a reduction in the frequency of nerve impulses in this system. In full light, widening of the pupil occurs during great emotional excitement. For instance, in the cat or monkey when confronted with an angry barking dog. This widening is mainly brought about by sympathetic excitation, but the shortening of the radial muscle fibres is facilitated by inhibition of the sphincter.

The second example concerns the heart: vagal stimulation slows the heart while sympathetic stimulation accelerates it. These two antagonistic systems exert a continuous control on the heart rate. In a state of complete rest the passage of impulses can be recorded both in the cardiac fibres of the vagus and in the sympathetic accelerator nerves which leave the stellate ganglia. During muscular exercise, it is mainly sympathetic stimulation which speeds up the heart, but cessation of vagal inhibition also contributes to the acceleration. At rest, the heart rate of athletes (particularly of swimmers and long distance runners) is much lower (40-50 per minute) than that of the non-athletic man ($\pm$ 72); we know that this slowing results from an increase in vagal inhibitory tone.

A third and last example: erection is due to sacral parasympathetic excitation, while ejaculation is controlled by the lumbar sympathetic, stimulation of which produces contraction of the vas deferens and the seminal vesicles, as well as secretion of the prostate. These organs have no antagonistic parasympathetic innervation. Certainly a fortunate arrangement since during coitus both systems are aroused simultaneously.

3

The forerunners*

No theory is tested experimentally until after a certain period of maturation. This becomes very obvious on reviewing the literature. Certain authors attribute great importance to a sentence by E. Du Bois Reymond, (dating back almost a century), who in his time was a prominent figure:

'Of known natural processes that might pass on excitation, only two are, in my opinion, worth talking about: either there exists at the boundary of the contractile substance a stimulatory secretion in the form of a thin layer of ammonia, lactic acid, or some other powerful stimulatory substance; or the phenomenon is electrical in nature.' (Gesammelte Abhandlungen der allgemeinen Muskel- und Nervenphysik, 1877, 2, 700).**

It was mainly the second hypothesis which formed the basis of research for physiologists at the end of the nineteenth and at the

---

* The reader will find a well-documented and illustrated study of this and the following two chapters in a review which I published in French in 1935 in Ergebnisse der Physiologie und experimentellen Pharmakologie (vol. 37, p. 82-185) at the request of Leon Asher and on the recommendation of W. B. Cannon. See also Dale, H. H., *Linacre Lecture*, 1934.

** 'Von bekannten Naturprocessen, welche nun noch die Erregung vermitteln könnten, kommen, soviel ich sehe, nur zwei in Frage. Entweder müßte an der Grenze der kontraktilen Substanz eine reizende Sekretion, in Gestalt etwa einer dünnen Schicht von Ammoniak oder Milchsäure oder einem anderen, den Muskel heftig erregenden Stoff stattfinden. Oder die Wirkung müßte electrisch sein.'

beginning of the twentieth century. It took half a century to discover the 'powerful stimulatory substance' of striated muscle.

At the beginning of this century (1906), Howell, an American physiologist, showed that even small changes in the cation concentration of saline solutions used for studying isolated organs produced effects similar to those which occur on nerve stimulation. An increase in the concentration of potassium ions in particular, slowed down and depressed the isolated heart; in addition, vagal stimulation increased the potassium content of the perfusing saline solution. Atropine, however, did not abolish the effects of potassium while it abolished those of muscarine and vagal stimulation. We may in fact say that Howell laid the first stone of another vast building, that of membrane biophysics of which Hodgkin and Huxley became the principal architects. We know today that the physiology and pharmacology of the heart, smooth muscle, glands and nerve cells are dependent on the existence of unequal concentrations of ions on both sides of the cell or nerve membrane and on very rapid changes in their concentrations which create the differences in the electric potentials which we record.

One of the first physiologists to express himself clearly on chemical transmission was T. R. Elliott, a young Research Fellow in Cambridge. In 1901, it was known that adrenaline, crystallized from an extract of the adrenal glands, reproduced the effects of sympathetic nerve stimulation, and that certain alkaloids extracted from ergot inhibited many actions of adrenaline and of sympathetic stimulation. In a short prophetic note presented to the Physiological Society on May 21, 1904, Elliott wrote 'Adrenaline might then be the chemical simulant liberated on each occasion when the impulse arrives at the periphery'. I can still recall how happy W. B. Cannon was when he brought me this note in December 1930; he was in the process of preparing our paper on sympathin for the *American Journal of Physiology*. Cannon loved to pay homage to the work of his predecessors or contemporaries with a generosity inherent to his nature. Two facts had struck Elliott. (1) That adrenaline reproduced the effects of sympathetic nerve stimulation, and (2) that degeneration of the post-ganglionic sympathetic fibres after removal of the ganglion cells did not abolish the sensitivity of the smooth muscle to this hormone, but on the contrary increased it. T. R. Elliott had drawn the logical conclusion.* W. E. Dixon (1907) who was familiar with the suggestion put forward by Elliott, attempted to demonstrate the release of a muscarinic substance on vagal stimulation in a dog's heart. The technical conditions of his experiments were poor. Since 1926 we have known that acetylcholine is very rapidly inactivated in mammals if one does not take the precaution of using eserine which inhibits the cholinesterases. Because of the universal scepticism of those around him, Dixon quickly abandoned his experiments, which could have taken him far along the right path. His writings as well as the testimonies of his friends show that he continually hesitated between the concept of chemical transmission and the ancient theory which claimed that mimetic poisons excite the nerve endings.** A paper published in the *Medical Magazine*, XVI, 454, 1907, shows how close Dixon was to the discovery of the key facts. His experiment represents the first attempt to extract Vagusstoff from the heart. Loewi himself stated in a film shown at the International Congress of Pharmacology (San Francisco, August 1972) that since 1913 he had thought about an experiment which he did not perform until much later.

---

* It is known, through his friend H. H. Dale, and through various writings, that Elliott never gave up the idea of chemical transmission. Feldberg and Fessard (1942) remind us that in 1914, Elliott, in his *Sidney Ringer Memorial Lecture* discussed the excitation of the teleost's electric organs under this aspect. He wrote 'I have tried in vain to discover an active substance in the muscle plates of striped muscles. . . . But it is hard to forego the belief that such discoveries lie in the lap of the future.' What a marvellous statement! Also, how tragic, to be unable to prove a concept that is so self-evident to oneself.

** This theory contained some truth, because for some time it has been known that indirect sympathomimetics (tyramine, for example) act, at least partly, by releasing a fraction of the noradrenaline stores from adrenergic nerve endings (J. H. Burn).

In their remarkable monograph on acetylcholine (1973), Michelson and Zeitmal gave more credit to J. N. Langley than is given to him by other authors, even of English nationality, who studied the origin and development of the theory which is the essence of this book. It seems that this interest in Langley originated in the U.S.S.R. from A.G. Genetsinsky. Langley's work on nicotine is universally known; all physiologists and pharmacologists acknowledge that he was the first to give a valid description of the idea of 'receptors', but no-one remembers that, in 1906, Langley had already taken the logical and decisive step which led him to postulate the existence of a 'chemical mediator'. The following quotation taken from Michelson and Zeitmal is very clear and is presented in a form that the molecular pharmacologists of today would understand: 'The stimuli passing the nerve can only affect the contractile molecule by the radical which combines with nicotine and curare. And this seems in its turn to require that the nervous impulse should not pass from nerve to muscle by an electrical discharge, but by the secretion of a special substance at the end of the nerve.' In the introduction to his important paper (1905) on the effects of adrenaline, Elliott did not forget to praise the work of Langley and to thank him for his advice. Why is there this relative neglect of this part of Langley's work? Is it deliberate? Had Langley annoyed his contemporaries by his excessively severe control of the *Journal of Physiology*? It does not matter now. Let us pay homage to his clear foresight as we have done for many years to T. R. Elliott.

Despite all this work, which did not go unnoticed by many outstanding scientists, half a century ago—around 1920—minds were still hardly prepared to accept the results and consequences first of the experiments of O. Loewi and then later of Dale and Cannon.

4

Cholinergic transmission

**4.1. First phase. The parasympathetic division of the auto-
nomic nervous system**

4.1.1. *The fundamental experiments of Loewi on the Amphibian
heart*

Stimulation of the vagus nerve causes a slowing of the heart
(Weber, 1845). In 1921, Loewi's fundamental experiment consisted
of replacing the normal salt solution inside a frog's heart
isolated according to the method of Straub, by the fluid pipetted
from another heart whose vagi had been electrically stimulated.
Numerous objections were raised, particularly against technical
aspects of this method. Undoubtedly, the cardiac muscle would
be disturbed by the transferring of the fluid from the one cannula
to the other. The truth is that in Loewi's first publication demonstrating the existence of 'Vagusstoff' the records were far from
convincing. Yet why did we have to wait five years before Kahn
in Prague removed these objections by using two isolated frog
hearts attached to a double branched cannula? Stimulation of the
vagus to one heart resulted after a long delay—150 seconds from
the beginning of stimulation—in a characteristic vagal effect on
the second heart, which had no contact with the first one except
through the saline solution (Fig. 3). Nowadays, confirmation of
Loewi's experiments would not have taken more than a few
months, and the technique of Kahn should have been developed
in Loewi's own laboratory.

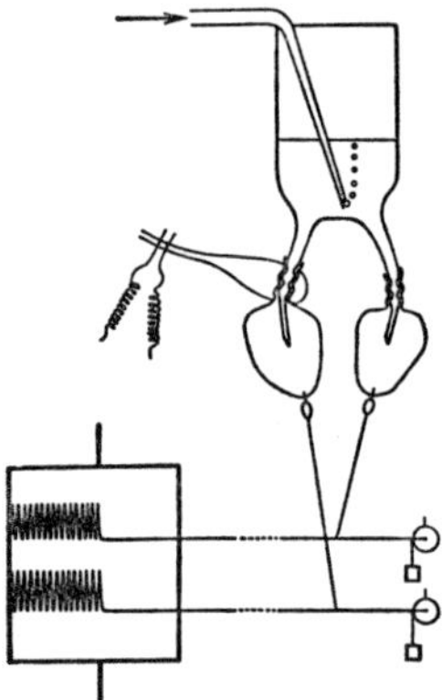

Fig. 3 *a*. — Double branched cannula for studying humoral transmission of vagus stimulation to the frog's heart. Two hearts are suspended on the two branches of the same cannula in such a way that they are bathed by the same fluid. If a vagomimetic substance is liberated on vagus stimulation of the donor heart, the other (receptor) heart should react to it. With this ingenious method any pipetting, as in Loewi's original famous experiment, is avoided. The objection therefore no longer applies that the effect is due to changes in hydraulic pressure.
(From Kahn, *Pflügers Arch. 214*, 485, 1926.)

Another objection was pointed out: the Amphibian heart is innervated not only by vagal inhibitory fibres, but also by fibres of sympathetic origin, stimulation of which accelerates the heart and augments the amplitude of the contractions. With the methods used by Loewi as well as by Kahn, both types of nerve were stimulated, but in their experiments stimulation usually resulted in cardiac slowing. The reason for this is related to seasonal changes in the sensitivity of the Amphibian heart. Stimulation of the vago-sympathetic nerve trunk leads to cardiac acceleration only in the spring, and Loewi made use of this phenomenon to demonstrate, in the spring of 1921, that the fluid which bathed the heart contained an 'Acceleransstoff'. In 1921, the opponents felt that they had some trumps in their hands, they

Fig. 3 *b*. — Tracings obtained with the method illustrated in figure 3 *a*. Top tracing contractions of the receptor heart; bottom tracing contractions of the donor heart. The signal above the bottom tracing indicates 140 seconds stimulation of the vagus to the donor heart which causes pronounced weakening and slowing of the heart beat. About 70 seconds after the end of this stimulation the contractions of the receptor heart become definitely weaker but not slower.
(From Kahn, *Pflügers Arch. 214*, 488, 1926.)

denounced Loewi's interpretation as being childish and over-simplified. This was asking too much! Here was an organ and we stimulated its nerves; if the effect was inhibitory, the substance liberated was inhibitory; if the effect was excitatory, the substance was an excitatory one. What magnificent simplicity! This kind of reasoning on the part of Loewi's opponents did not fail to make its impression, but Loewi stuck to his views.

The choice of the frog heart was a fortunate one. We know that in warm- or cold-blooded Vertebrates, electrical stimulation of the cut peripheral end of the vagus (cut in order to avoid central effects) results in slowing of the heart and even in cardiac arrest as well as in a decrease in the force of the contractions. Since 1867, small doses of atropine (an alkaloid of belladonna) had been known to abolish these effects of vagal stimulation. These facts were difficult, if not impossible, to explain by the classical electrical theory prevalent at the time. Since the beginning of the 20th century it had been known that electric currents pass along the nerves and skeletal striated muscles when they were stimulated. It was generally accepted that the electrical variations which passed along the nerve fibre arriving at the nerve ending would be able to pass (by a simple physical mechanism) to the membrane of the muscle which showed under the microscope a specialized structure, the motor end plate.

About half a century ago, when I was a student, the inadequacy of this theory was strongly felt in the case of stimulation of the cardiac vagus. Even if one could imagine an electric current arriving at the motor end plate of a striated muscle to trigger an impulse along the length of the muscle fibre, it was difficult to see how this same current could inhibit or stop automatic activity at the region of the vagal endings. The inadequacies of the classical theory during this era were flagrant. Many details remained unexplained, such as the delay in the disappearance of the effects of vagal excitation after the end of a period of electrical stimulation of the nerve, especially in a cold blooded animal such as the tortoise.

In their traditional arsenal, pharmacologists possessed a substance called muscarine, extracted from *Amanita muscaria,* which was poorly defined at the time. It inhibited or diminished the force of the cardiac contractions, caused slowing and then arrest of the hearts of Vertebrates; the effects of muscarine were abolished by atropine. Again, the interpretation of these facts by the classical theory could satisfy conventional minds, but not those of exacting physiologists. It was said, without giving any proof, that muscarine reproduces the effects of vagal stimulation because it excites these nerve endings. The lack of logic, the weakness of this interpretation, should have aroused the imagination of pharmacologists and should have instigated experiments long before Loewi published his decisive experiments. W. E. Dixon made the fundamental mistake of assuming that muscarine acts by the release of an inhibitory hormone; this misinterpretation led him to the further error of suggesting that atropine inhibits the release of this hormone instead of occupying the receptor.

In a short, but attractive biographical sketch (1960) * Loewi told the story of two dreams on successive nights during Easter of 1920 which recalled to him an idea he had in 1903 during a discussion with W. M. Fletcher of Cambridge (U. K.). He gave an explanation of the long delay which did not entirely convince me, in as much as a parallel concept had evolved in 1904 in the domain of the sympathetic system when Elliott proposed his ingenious idea which was treated with a great deal of reservation by Barger and Dale in 1910 (see page 93 of *Adventures in Physiology* by H. H. Dale).

'As far back as 1903, I discussed with Walter M. Fletcher from Cambridge England, then an associate in Marburg, the fact that certain drugs mimic the augmentary as well as the inhibitory effects of the stimulation of sympathetic and/or parasympathetic nerves on their effector organs. During this discussion, the idea occurred to me

---

* Original text: An Autobiographic Sketch, pp. 17 and 18. *Perspectives in Biology and Medicine,* IV, No. 1, Chicago, 1960.

that the terminals of those nerves might contain chemicals, that stimulation might liberate them from the nerve terminals, and that these chemicals might in turn transmit the nervous impulse to their respective effector organs. At that time I did not see a way to prove the correctness of this hunch, and it entirely slipped my conscious memory until it emerged again in 1920.

The night before Easter Sunday of that year I awoke, turned on the light, and jotted down a few notes on a tiny slip of thin paper. Then I fell asleep again. It occurred to me at six o'clock in the morning that during the night I had written down something most important, but I was unable to decipher the scrawl. The next night, at three o'clock, the idea returned. It was the design of an experiment to determine whether or not the hypothesis of chemical transmission that I had uttered seventeen years ago was correct. I got up immediately, went to the laboratory, and performed a simple experiment on a frog heart according to the nocturnal design. I have to describe briefly this experiment since its results became the foundation of the theory of chemical transmission of the nervous impulse.

The hearts of two frogs were isolated, the first with its nerves, the second without. Both hearts were attached to Straub cannulas filled with a little Ringer solution. The vagus nerve of the first heart was stimulated for a few minutes. Then the Ringer solution that had been in the first heart during the stimulation of the vagus was transferred to the second heart. It slowed and its beats diminished just as if its vagus had been stimulated. Similarly, when the accelerator nerve was stimulated and the Ringer from this period transferred, the second heart speeded up and its beats increased. * These results unequivocally proved that the nerves do not influence the heart directly but liberate from their terminals specific chemical substances which, in their turn, cause the well-known modifications of the function of the heart characteristics of the stimulation of its nerves.

The story of this discovery shows that an idea may sleep for decades in the unconscious mind and then suddenly return. Further,

---

* Note from the Author: At the time when Loewi wrote this story (1960 at the age of 80) he forgot that with his technique he always stimulated simultaneously both vagal inhibitory fibres and sympathetic fibres. It is possible but difficult to separate the two kinds of nerves.

it indicates that we should sometimes trust a sudden intuition without too much skepticism. If carefully considered in the daytime, I would undoubtedly have rejected the kind of experiment I performed. It would have seemed likely that any transmitting agent released by a nervous impulse would be in an amount sufficient to influence the effector organ. It would seem improbable that an excess that could be detected would escape into the fluid which filled the heart. It was good fortune that at the moment of the hunch I did not think but acted immediately.

For many years this nocturnal emergence of the design of the crucial experiment to check the validity of a hypothesis uttered seventeen years before was a complete mystery. My interest in that problem was revived about five years ago by a discussion with the late Ernest Kris, a leading psychoanalyst. A short time later I had to write my bibliography, and glanced over all the papers published from my laboratory. I came across two studies made about two years before the arrival of the nocturnal design in which, also in search of a substance given off from the heart, I had applied the technique used in 1920. This experience, in my opinion, was an essential preparation for the idea of the finished design. In fact, the nocturnal concept represented a sudden association of the hypothesis of 1903 with the method tested not long before in other experiments. Most so-called 'intuitive' discoveries are such associations suddenly made in the unconscious mind.'

One can accept or reject O. Loewi's story. However, it is certain that genetically well organized and active minds automatically select facts, names, figures and ideas and retain them. One must know how to forget what is useless!

The young biologist of today should realize that this delay in the propagation of a fruitful hypothesis was mainly due to the fact that at the beginning of our century the number of biologists and pharmacologists was very small, meetings and congresses were infrequent, journeys were long and difficult, and the development of the physical and chemical sciences on which biology depends was also slow. As I see it, the main change that has occurred since then is the great expansion of the scientific community.

If a statistical analysis were possible it would show that in this community, the percentage of open minds and of dynamic characters capable of taking the great scientific steps is of the same magnitude in 1972 as it was in 1936 or in 1900. It is only natural that even the best may hesitate and take the wrong road. In 1950, although the triumph of the chemical transmission theory was undeniable, certain eminent physiologists still refused to incorporate this theory into their teaching.

The observation which provided Loewi with his principal weapon was his discovery in 1926, with Navratil, of the inhibitory action of eserine (or physostigmine, an alkaloid of the Calabar bean, isolated in 1864) on the esterase, the enzyme which inactivates the Vagusstoff already suspected to be acetylcholine or a closely allied substance.

From then on, one had not only a working hypothesis, but also the possibility of regularly reproducing the various decisive experiments, and not exclusively on Amphibians, but also on Mammals. Looking back, one must regard the choice of the Amphibian heart by Loewi as being providential. For instance, the cholinesterase content in this organ is small compared to that in mammalian hearts, the low temperature is favourable for the stability of acetylcholine, and working in a saline medium avoids the cholinesterase in the erythrocytes and plasma. Loewi had unwittingly combined the most favourable conditions to ensure the success of his experiments. The great merit of this exceptional man was his ability to follow a logical course and to accumulate convincing experimental evidence without paying too much attention to the criticisms of his theory. The main thing is to have confidence in one's own view of the truth.

Between 1926 and 1936, and particularly after 1933, a constant stream of outstanding experiments was produced, among them those of Dale and his associates who played such a prominent role. In 1933, in Berlin, Feldberg and Krayer showed in a faultless way, with all necessary controls, that vagal stimulation of the dog heart liberated into the blood, an acetylcholine-like substance which contracted the leech muscle and lowered the cat's blood pressure, provided the esterase was inactivated by eserine.

4.1.2. *Extension of Loewi's concept to the whole parasympathetic innervation in Mammals.*

After 1932, the principal methods became well established and were used for about ten years, the time needed by the chemists to develop sufficiently specific and sensitive micromethods. It is to Feldberg that we owe the systematic introduction of eserine to inhibit the cholinesterase, and a curious test: the dorsal muscle of the Hungarian leech, *Hirudo officinalis,* discovered by Fühner. B. Minz, on Feldberg's suggestion, examined systematically the pharmacology of this muscle, and proved that in extremely low concentrations acetylcholine was practically the only physiological substance present in blood or saline perfusates which would produce contractions.*

The preparation in the anaesthetized animal (dog, cat or rabbit) consists of placing ligatures and cannulae in such a way as to isolate the circulation of an organ and then to collect at a chosen time the blood or a saline perfusate which has passed through the organ. If the eserinized leech muscle contracted to the fluid collected during stimulation of the peripheral end of the cut nerve, the next step was to show that treatment of the leech muscle with a specific antagonist of acetylcholine (in this case curare) prevents or abolishes the contraction. After the fundamentals had been established, skill, patience and the will to succeed made it possible to produce simple records and convincing tracings.

---

* Chang and Gaddum showed in 1933 that for the biological assay of acetylcholine in tissue extracts, the *rectus abdominus* of the frog is preferable to the leech muscle.

Until 1933, W. Feldberg worked at the Institute of Physiology in Berlin, whose director was Wilhelm Trendelenburg. In 1925, he took his MD and then went to England for two years to work with J. N. Langley, J. Barcroft and H. H. Dale; in 1930 he became Privat Dozent. In 1932, at the International Congress of Physiology in Rome, he met O. Krayer from the Institute of Pharmacology in Berlin and they decided to work together. Krayer was an expert in the heart-lung preparation of the dog and in the collection of blood from the coronary sinus. Feldberg had previously successfully applied his technique to the detection of acetylcholine in the venous blood from the dog's tongue during electrical stimulation of its parasympathetic nerves—the chorda tympani—which dilates its vessels. In the distressing atmosphere of 1933—the year when Hitler seized power in Germany—Feldberg and Krayer demonstrated in an elegant way the release of an acetylcholine-like substance from the heart of a dog during stimulation of its vagal nerve (Fig. 4).

This was the moment when Feldberg had to emigrate and joined Dale.* A little later O. Krayer also came to England and worked for a time with Verney, he then moved first to Beirut and finally settled at Harvard as Head of the Department of

---

* Dale, who knew well how to choose his associates, took him into his famous group. Those who knew the exciting atmosphere of the years 1933-39, and worked in Room F4 of the National Institute for Medical Research, formed a sort of spiritual family and remained united despite dispersion. Alas, many of them have passed away. In 1974, Feldberg is still active doing original research, particularly in the field of the central nervous system. He is always introducing new techniques and new interpretations. One meets him at Congresses in Great Britain and abroad with his young co-workers whom he trains with friendly severity. No pharmacological physiologist has been as successful as he in gathering around him a group of co-workers and students who have nothing but the highest praise for him. For those who have the pleasure of knowing him, he remains a devoted friend, a broadminded colleague, an amazing master, always accessible, for whom research is the essence of life. He is 'possessed' by the demon of scientific research. Without Feldberg, the research on chemical transmission of nerve impulses might well have taken a very different course.

Pharmacology. Many other German physiologists, biochemists and pharmacologists left their country during the years 1933-35. Among them were: Marthe Vogt, Edith Bülbring, Blaschko, Nachmansohn and Minz. What a blood-letting of talent on the one hand, and what a precious enrichment on the other! The contributions made in Great Britain and the U. S. in the field of chemical transmission, the history of which is told in these pages, would not have been nearly as impressive without the contributions by this group of refugees.

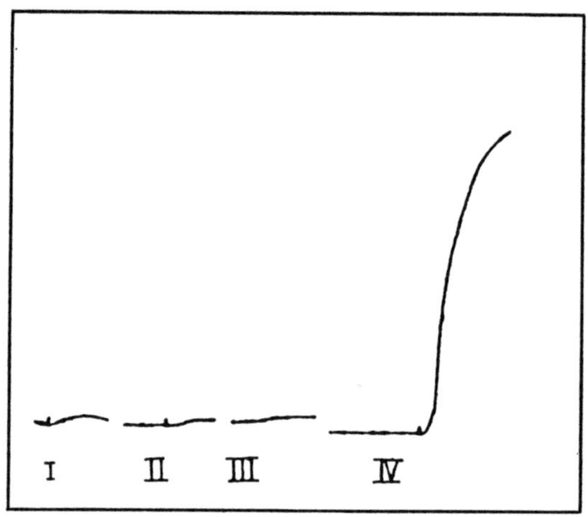

FIG. 4 — Effect on the eserinised leech muscle preparation of blood coming from the coronary sinus of the dog's heart. As long as the dog had not been eserinised neither the blood before (I) nor that during (II) vagus stimulation produces contraction. Any 'Vagusstoff' released on stimulation would immediately be inactivated by the cholinesterase. After the dog has been given an intravenous injection of 10 mg of eserine to inactivate the cholinesterase the blood coming from the heart vein without vagus stimulation is again inactive (III) but that coming during vagus stimulation produces a strong contraction (IV). This experiment proves that vagus stimulation releases an acetylcholine-like substance into the blood which, in the absence of eserine, is at once inactivated.

(From Feldberg & Krayer, *Arch. exp. Path. Pharmacol. 172*, 177, 1933.)

After his arrival in London, Feldberg worked with Dale and they proved that on vagal stimulation acetylcholine appeared in the venous blood of the stomach of a dog. They not only used Feldberg's favourite test, the leech muscle, but also the isolated heart, the *rectus abdominis* muscle of the frog and in addition the arterial blood pressure of the cat (Fig. 5). Various authors demonstrated that excitation of the parasympathetic nerves to the eye (which contract the pupil), to the salivary glands and the uro-genital organs release a substance which has the same properties as Loewi's 'Vagusstoff'.

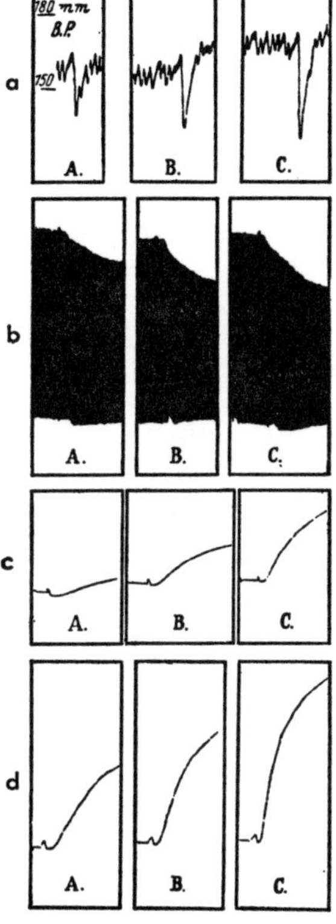

FIG. 5. — Effect of concentrated effluent (eserinised Locke's solution) obtained from the perfused pyloric part of the dog's stomach during vagus stimulation on (a) arterial blood pressure of a cat under chloralose anaesthesia, (b) isolated frog heart (Straub), (c) eserinised frog rectus muscle and (d) eserinised leech muscle. In all three preparations B shows the effect of concentrated effluent in a constant ratio to two strengths (A and C) of acetylcholine solutions, that of C being double that of A. (Concentrate of control effluent obtained before vagus stimulation had no or little effect on the three preparations.) This figure shows that the effect of the acetylcholine-like transmitter in the effluent resembles qualitatively and quantitatively that of acetylcholine, its effect being approximately mid-way between that of A and C on all three preparations.

(From Dale and Feldberg, *J. Physiol. 81*, 331, 1934.)

4.1.3. *Problems of nomenclature*

In 1933, it was well established that organ extracts of Vertebrates contained appreciable amounts of a substance whose chemical properties and physiological effects were identical with those of acetylcholine. In 1929, Dale and Dudley had isolated acetylcholine and identified it chemically from an extract of horse spleen. It was shown that certain nerve fibres which belong to the 'parasympathetic' division of the autonomic nervous system released acetylcholine on stimulation of the fibres which run in the IIIrd cranial nerves, in the chorda tympani, in the Xth cranial nerves, the vagi, and in the sacral nerves to the bladder and the genital organs.

It did not take long before it was discovered that there were also endings of postganglionic fibres belonging to the 'sympathetic' division, the thoracolumbar outflow (leaving the spinal cord between the first dorsal and the third lumbar vertebrae) which also release acetylcholine in certain animals, for example, the nerve fibres to the sweat glands in the paw of the cat. Although being 'anatomically' sympathetic, a nerve fibre may have the biochemical properties of synthesizing, transporting and releasing not an 'adrenaline-like sympathin' but acetylcholine. Furthermore, the preganglionic fibres of the sympathetic division which leave the spinal cord to end in sympathetic ganglia release acetylcholine like the motor nerves of skeletal muscles. It thus became necessary to superimpose a physiological nomenclature onto the anatomical one, to indicate the nature of the substance synthesized and released at the nerve endings. It was Sir Henry Dale who in 1933 proposed that those fibres which liberate acetylcholine be called *cholinergic* and those which use adrenaline or an allied substance as transmitter, *adrenergic*. Dale and his co-workers suggested that the term *cholinergic* be applied only to nerve fibres which act through the release of a choline ester. A receptor or an allied substance may be called *cholinoceptive* or *cholinomimetic* respectively but not cholinergic. Dale fought hard to prevent the terminology he had coined from being used for purposes other than those for which it was intended. But the terms were soon wrongly applied and used to characterize all kinds of substances and all kinds of phenomena related to the functioning of nerve fibres. The misuse today is so widespread that we do not even recognize the unconscious insult perpetrated on his elegant concept which, at the time, he made every effort to define.

Of more recent date is the use, particularly in the pharmacology of the central nervous system of the term *tryptaminergic* (or serotoninergic) for nerve fibres which release 5-hydroxytrypta-mine (5HT) and the term *monoaminergic* for nerve fibres which release either a catecholamine or 5HT. The term *dopaminergic* applies to neurons that synthesize and release dopamine without converting this amine into noradrenaline.

American authors use the terms *epinephrine* for adrenaline, and *norepinephrine* for noradrenaline. In W. B. Cannon's time the term *adrenin* was used to define the substance or substances secreted by the adrenal glands. Today we know that these sub-stance are adrenaline and noradrenaline (rarely dopamine) and that they occur in different proportions according to the species, and that in each species they are secreted in different proportions according to the physiological requirements (see also p. 36 and 54).

4.2. SECOND PHASE. THE CHOLINERGIC NATURE OF THE MOTOR
     NERVES TO STRIATED MUSCLES AND OF THE PREGANGLIONIC
     FIBRES OF THE AUTONOMIC NERVOUS SYSTEM

Until 1932 all experiments which dealt with cholinergic or adrenergic transmission had been on peripheral organs such as the heart, smooth muscle and glands which receive nerve fibres from the autonomic nervous system. The known facts about the general physiology of these organs were in full agree-ment with the concept of chemical transmission, whether the nerves were cholinergic (the two sections of the parasympathetic division) or whether they were adrenergic and belonged to the sympathetic outflow from the spinal cord which spreads all over the body. In particular, there is the long delay between electrical stimulation of the nerve and the onset of the response of the innervated organ; the response itself is long lasting and often persists after the nerve stimulation has come to an end, and there is also a very gradual return of the organ to its basal activity. One needs only to look at an experiment in which autonomic nerves are stimulated to become convinced that by far the best general theory which would account for all the observed facts is the one which allows a more or less labile substance to be interposed between the nerve ending and the effector cell, be it a smooth muscle or a gland cell.

The response of striated muscle to motor nerve stimulation had been widely studied. In particular, the changes in the electrical potential in the nerve and muscle fibres had been recorded. It was known that a single stimulus or a single wave of depolarisation in the nerve was followed, after a short latency, by a single response in the muscle, the nature of which was first electrical and then mechanical. No-one questioned the concept that the conduction of the nerve impulse along the nerve fibre and the transmission of excitation to the muscle was electrical (physical). Yet there were numerous pharmacological observations which could not be interpreted readily with this over-simplified theory. In the first place, there was the effect of curare. Claude Bernard had demonstrated unequivocally that curare—he had obtained a sample of South American curare from Napoleon III— did not abolish the ability of the muscle to contract, nor did it injure the nerve. The site at which transmission was blocked was the region called the motor end plate, where the nerve fibres end and where the histologists had described very specialized structures. It was also known that in the sympathetic nervous system the same curare would prevent impulses in the pre-ganglionic nerve fibres from being transmitted to the ganglion cells which, from the anatomical point of view, are experimentally more easily accessible.

It had also been known for a long time that acetylcholine contracted the isolated striated muscle of frogs and toads suspended in a bath of physiological saline solution and that the contraction was abolished by curare and atropine. Then there was the wonderful study by Langley of the action of nicotine (an alkaloid of tobacco) on striated muscle and on ganglion cells. He observed that this poison first excited these structures and then blocked transmission. It was also known that an intravenous injection of acetylcholine into a mammal (protected by atropine) resulted in a rise in arterial blood pressure which since 1934 is generally thought to be due to stimulation of ganglion cells, the axons of which innervate the muscle fibres of the arterioles.

In the same year (1933), and in the same volume (232) of the Pflügers Archiv, the classical periodical of the German physiologists, two papers of paramount importance appeared.

(1) W. Feldberg, still at the Institute of Physiology in Berlin, made use of the favourable anatomical arrangement of the innervation of the mammalian tongue, where the motor nerve is fully separated from the parasympathetic cholinergic vasodilator fibres of the chorda tympani which emerge from the lingual nerve and from the sympathetic fibres which originate in the superior cervical ganglion and run in the peri-arterial connective tissue sheath. Feldberg perfused half of a dog's tongue with blood and, with the help of suitably placed ligatures, was able to collect all the venous blood whenever he wanted, and to pass it through a tube into a bath in which a leech muscle was suspended. The dog was eserinised to inhibit the cholinesterase, and atropinised to prevent the cardiodepressor effects of eserine. Stimulation of the lingual nerve caused the appearance of a substance in the venous blood which contracted the leech muscle (Fig. 6). This substance was destroyed by blood in the absence of eserine and, like acetylcholine, it produced a fall in arterial blood pressure in the cat which was abolished by atropine. Feldberg reported this observation briefly at the Rome Congress in 1932.

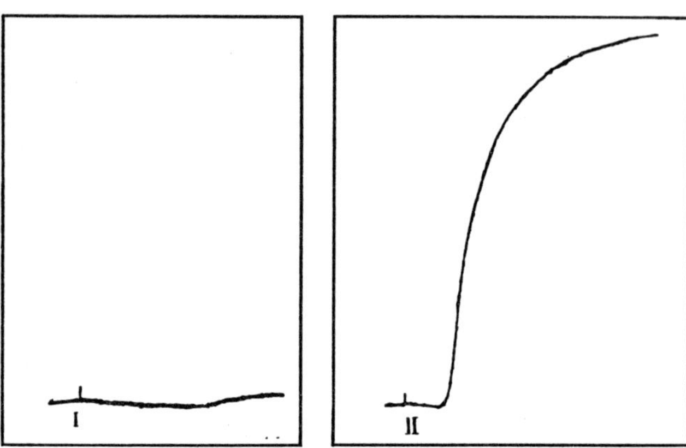

FIG. 6. — Effect on the eserinised leech muscle preparation of eserinised blood coming from the dog's tongue before (I) and during (II) stimulation of the lingual nerve.
(From Feldberg, *Pflügers Arch. 232*, 93, 1932.)

(2) Kibjakow, in Kazan (U.S.S.R.) was the first to succeed in perfusing the superior cervical ganglion in the cat. The principle of the perfusion was nothing new. A cannula was inserted into the large artery which passes along the region of the organ to be perfused (in this case the artery was the common carotid). Then one patiently tied off all the vessels which did not go into or come from the ganglion. The venous blood was collected from the jugular vein. A physiological salt solution which was adequate to maintain the functioning of the ganglion cells, not blood, was used for the perfusion. If one collected the perfusate during stimulation of the preganglionic nerve (the cervical sympathetic) and injected it either into the same preparation or into another perfused ganglion, one observed a contraction of the nictitating membrane. This smooth muscle receives postganglionic fibres which originate from the superior cervical ganglion. One could therefore conclude that the stimulation of the preganglionic nerve fibres had released at their endings a substance which was

able to stimulate the ganglion cells. Kibjakow was fully aware of the importance of his experiment. For the first time in the history of physiology it was demonstrated that transmission from one neurone to another could be chemical in nature. Kibjakow did not use eserine in his experiment. He did not consider the possibility that the substance might be acetylcholine, although he should have been aware that Dale and Gaddum (1930) had suggested an acetylcholine-like substance as mediator of the parasympathetic vasodilator effects. Kibjakow gave a demonstration of his experiment at the International Congress of Physiology (Moscow-Leningrad, 1935) and afterwards a photograph was taken (Fig. 7). Sir Henry Dale (Fig. 8) is missing from this family picture; he never liked the U.S.S.R.

4.2.1. *Striated muscle*

It was in London, at the N.I.M.R. (National Institute for Medical Research) in Hampstead, under Dale's direction, that this aspect of the research came to fruition, thanks to W. Feldberg, G. L. Brown, J. H. Gaddum and Marthe Vogt.

Let us first take a glance at the history of research on neuromuscular transmission. In 1933, the evidence concerning chemical transmission was very meagre. It was known that enormous doses of acetylcholine had to be injected in order to produce a contraction of normal mammalian muscle. Feldberg in Berlin, and Simonart in Belgium had obtained some very suggestive evidence. The theory of chemical transmission postulates that the transmitter must reproduce the effects of nerve stimulation. An ingenious technique, of which G. L. Brown became the unsurpassed master, made it possible to obtain from the cat *gastrocnemius* muscle (and later from the *anterior tibialis* muscle) tensions of several kilograms following rapid close-arterial injection of a few micrograms of acetylcholine. Under optimal experimental conditions a dose of only 0.25 μg was effective. Eserine potentiated the effects of motor nerve stimulation, provided single stimuli were applied, or the frequency of stimulation was low (six per minute or less) even if stimulation was supramaximal, (i.e., if all fibres of the motor nerve became excited and a single wave of depolarisation was produced). If, a tetanus was applied for about twenty seconds after eserinisation, the contraction was not maintained, and if stimulation with single shocks was then resumed the response was reduced in comparison to that recorded before the tetanus. The reason is that if acetylcholine is not immediately hydrolyzed and inactivated by the cholinesterase and persists at the synapse, it exerts a curare-like action. This fact, which at first seemed paradoxical, was not accepted by Eccles, who elaborated many hypotheses to explain the double action of eserine There was another objection! Since the beautiful demonstration of Kato in Moscow, physiologists knew that according to the 'all or none' law, the response of a striated muscle to stimulation of an isolated nerve fibre was always maximal. G. L. Brown, who was also a good electrophysiologist, showed that after eserinisation, the action potential of a striated muscle was no longer a single deflection but consisted of a short series of 'well synchronized' action potentials. In short, the eserinized muscle responds to a single shock with a brief tetanus; acetylcholine when it persists for a short time at the motor endplate produces repetitive excitation. And the essential control! If we use a muscle which has been denervated six days previously, all the effects of eserine are absent; yet acetylcholine is now much more potent, and in addition it depresses the effects of direct electrical stimulation. The whole pharmacology of striated muscle takes on a different appearance in the light of the concept of 'cholinergic' transmission. Curare acts on the motor endplate on what are now called cholinergic receptors, and the receptors become insensitive to the acetylcholine released on nerve stimulation. One can distinguish between two classes of curarizing substances according to whether they are depolarizing or non-depolarizing on the motor endplate. The biophysicists enter the field (Katz, Kuffler, etc.); Paton gives an ingenious theoretical interpretation of the phenomena of curarisation. G. L. Brown becomes an outstanding expert on the physiology and pharmacology of striated muscle. With von Euler he studied the post-tetanic potentiation, and with A. M. Harvey

Fig. 7. — Photograph taken in the laboratory of Professor Bykov in Leningrad during the XIVth International Congress of Physiology, 1935.
From left to right, seated in the first row: Galperin, S. I., Feldberg, W., Cannon, W. B., Bacq, Z. M., Bykov, K. M.; standing in the second row: Michelson, M. S., Epstein, Ya. A., Vladimirov, G. E., Andreev, L. A., Kibjakov, A. V., Gaddum, J. H., Shkolnikova, R., Solovjev, A. V., Olnianskaya, R. P., Rickl, A. V., Vasiljev, M. E., Goldenberg, E. E.

Fɪɢ. 8. — Sir Henry Hallet Dale.

the peculiarities of the eye muscles and of the muscles in birds and in a myotonic goat. During the war he worked successfully for the British Navy, but after the war he continued his previous work on muscle physiology at Hampstead with his many friends, W. Feldberg, B. D. Burns, E. Bülbring, J. A. B. Gray, M. Goffart and M. Vianna Dias.

In 1936, Brown and I showed that miotine and other anti-cholinesterases acted like eserine; we swept away the last objections to the chemical theory. After that time, J. C. Eccles remained almost the only important physiologist who did not admit that all we knew about the physiology and pharmacology of mammalian striated muscle was readily explained if one accepted the idea that upon the arrival of a nerve impulse at the ending of a motor axon (i.e. in the presynaptic structures) a definite number of preformed acetylcholine molecules are released, cross the short gap separating the pre- and postsynaptic membranes, excite the specialized receptors and are immediately inactivated.* Feldberg, always ready to accept a challenge, did not hesitate to join the French electrophysiologist, Fessard, at Arcachon just before the beginning of the last World War to prove that the nerves of the electric organ of *Torpedo* (an electric fish) were cholinergic and that acetylcholine was electrogenic (*J. Physiol.*, 1942, *101*, 200). Embryologists have shown that the electric organs of fishes originate from the same structures as striated muscle. After the war it became possible, thanks to the immense technical progress in electron microscopy and histochemistry, to see the storage vesicles in the presynaptic structures and the incredible concentration of acetylcholinesterase in the

---

* Of course, there remain a few facts which cannot be explained logically, and there has been a tendency to forget about them. For example, I showed Dale that eserine does not affect the responses to motor nerve stimulation in the frog, although the *rectus abdominus* muscle of this animal is used as the classical test preparation for assaying acetylcholine; furthermore atropine abolishes the acetylcholine contraction of this muscle but does not prevent the contraction produced by stimulation of its motor nerve.

electric plates of these organs. (See, for example, the monograph of E. de Robertis: *Histophysiologie des synapses et Neurosécrétion*, Gauthier Villars, Paris, 1964).

4.2.2. *Preganglionic fibres*

The study of transmission in the ganglia of the autonomic nervous system went on in parallel with that of striated muscle. At the end of 1933, the really beautiful experiments of Feldberg and Gaddum, which were in some ways a repetition of those of Kibjakow, showed that he had been right, but moreover that the substance responsible was acetycholine. The acetylcholine content of extracts of sympathetic ganglia is very much higher than that of striated muscle. Dale said that the ganglion can be looked upon as an accumulation of densely packed motor end plates, and that the acetylcholine is not diluted as in muscle extracts by the large masses of protein. After cutting the preganglionic sympathetic nerves the acetylcholine stores in the ganglion become gradually depleted. MacIntosh in London, and Coppée and I in Liège, demonstrated that a few days after sectioning when the store had fallen to about 25 % of its normal level, transmission of the impulse no longer took place. Curare blocks transmission, but as in the case of striated muscle, it does not prevent the excitation of the nerve endings from releasing the preformed acetylcholine molecules. MacIntosh and Brown studied the synthesis of acetylcholine in the perfused superior cervical ganglion. Synthesis is very active in this small collection of nervous cells. Even a long stimulation will not exhaust the acetylcholine store provided one does not forget to add to the perfusion fluid glucose, the preferred energy substrate, and choline, the immediate precursor.

Thus, the experiments of Langley which showed the similarity of the reaction of striated muscle and ganglionic cell to nicotine could be logically interpreted. On the other hand, neither the motor end plates of skeletal muscle nor the ganglion cells were excited by muscarine. The extremely fruitful concept of two types of receptor sensitive to and activated by acetylcholine is based on these findings, and Dale was again its originator. The two types of receptor are (1) the muscarinic receptor (heart, smooth muscle, salivary glands, etc.) excited by muscarine and allied synthetic substances; paralysed by atropine and atropine-like drugs; (2) the nicotinic receptor (striated muscle, ganglion) sensitive to many substances structurally related to nicotine and paralysed by curarizing drugs.

The receptors in different striated muscles and ganglia are not exactly the same. There are also species differences, and in the same species, differences between one muscle and another. Red striated muscles do not react in the same way as the white ones. Rat muscle does not react like dog or cat muscle. There are also great quantitative differences. Hexamethonium paralyses the ganglia in doses which do not affect striated muscles.

The well established ideas for the peripheral systems do not apply to the central nervous system without several important modifications. Atropine appears to be the most active agent. Curare and curare-like substances, which do not cross the blood-brain barrier have, however, unexpected side effects when injected into the cerebral ventricles: they then cause excitation and epileptic convulsions.

5

Adrenergic transmission

Since Loewi's discovery of the 'Acceleransstoff' in 1925, many observations have been published confirming the release of an adrenaline-like substance on stimulation of the sympathetic cardiac nerves (Henri Fredericq at Liège already in 1925; Brinkman and Van Damm, as well as Lanz in the Netherlands; Azler and Müller, Külz in Germany, Rijlant at Brussels). In 1930, Finkleman demonstrated the transmission of the inhibitory effects of sympathetic stimulation from one intestinal loop to another. Lehman saw, in 1932, that the perfusate from a hind leg of a frog collected during abdominal sympathetic stimulation, acted on the isolated heart like a diluted solution of adrenaline. Bain observed in 1933, a transitory inhibition of the isolated intestinal preparation when the bath fluid was replaced by perfusate collected from the perfused tongue of a dog during stimulation of the sympathetic vasoconstrictor nerves. There was, however, no experimental evidence that these observations were not artefacts, and that the same facts could be demonstrated in the intact mammal under normal physiological conditions.

This evidence was obtained by W. B. Cannon and his associates in 1930-31. I took part in this work due to the following circumstances: When I arrived at Harvard in September 1929, thanks to a grant from the Commission for Relief in Belgium (later to become the Belgo-American Education Foundation, B.A.E.F.), my scientific background was very poor. I knew practically nothing about the physiology of the autonomic nervous system.

I had spent a very happy year with André Mayer at the Collège de France. It was, in fact, André Mayer who sent me to W. B. Cannon. By good fortune, W. B. Cannon was looking for a co-worker to tackle a problem which had been growing slowly in his laboratory. I was available, I liked experimental surgery, I had been well trained by the surgeons at the University of Brussels, and I knew that I was not clumsy with my hands. The technique chosen by Cannon required the aseptic removal of the sympathetic fibres innervating the cat's heart about one week before the actual experiment, so as to exclude all direct nervous interference and to sensitize the heart to adrenaline. During the autumn of 1929 I prepared several cats for W. B. Cannon, who was well satisfied with the condition of the animals, as well as with the outcome of the first experiments. Cannon gave a large number of lectures in Europe in 1930; he met Dale, and discussed with him his views about his recent unpublished experiments. When he returned he was not in good health. He had to limit his activities and decided to carry out his research with me because I could relieve him of much of the heavy physical burden of the work. I prepared the animals, I also performed the spinal cord transections under ether anaesthesia, and then Cannon undertook the final preparations and carried out the experiment with me, which he did with wonderful precision and elegance. It was also he who took care of the recording on the smoked paper of a large kymograph, by which we measured cardiac frequency and registered arterial blood pressure. Sometimes when he was tired and the animal was still in good condition, I completed the experiment alone. We spent hundreds of hours working together in an atmosphere of complete absorption. The American physiologists on Cannon's staff were jealous of me because I spent so much time with their 'boss'. However, I did not realize this until just before my return in December of 1930, so great had been my pleasure to participate at first hand in such an important piece of research.

In the course of his work on sympathectomy in cats, Cannon had observed that after denervation of the heart, removal of one adrenal gland and denervation of the other, a mild cardiac acceleration was still obtained when the animal was submitted to an emotional stress. This was a weak progressive acceleration, reaching a maximum only three minutes after the beginning of excitation and disappearing after removal of the two abdominal sympathetic chains and the remnants of the thoracic chains, i.e., after disconnecting all peripheral ganglia from the central nervous system. In 1921 and 1922, Cannon had observed with Uridil and Griffith that after removal of the adrenals, excitation of the peripheral end of the nerve passing along the hepatic artery caused the liberation of a cardio-accelerator and hypertensive substance into the blood. They had called it 'liver substance'. The hypothesis which circulated at Harvard early in 1929, was that a substance released at the endings of functioning sympathetic fibres would pass into the blood and act on the heart very much like adrenaline. This hypothesis was verified by Cannon and myself, by stimulating the two abdominal sympathetic chains for 30 seconds in cats which had been prepared as I described above. We showed: (1) that the cardio-accelerator substance passed into the blood, by temporarily stopping the circulation in the aorta and vena cava at various times in relation to the stimulation; (2) that it was not only cardio-accelerator but also vaso-constrictor, and that it stimulated salivary secretion (Fig. 9) and (3) that ergotamine increased the amount of 'sympathin' released during stimulation.

When, several days before my departure, W. B. Cannon brought me the sheets of paper which were covered with his calm, regular handwriting without any alterations, I realized that he was pleased. He had discovered Elliott's note to the Physiological Society (1904) and he told me of the pleasure which he had had in the writing of the paper, the account of the facts, and above all in the discussion, where his great experience and his own particular style worked marvels. In my mind, this manuscript was perfect, even today one could not improve any part of it. I could easily have remained at Harvard University, and made my career in the United States; I knew that my master,

W. B. Cannon, was an exceptional personality, unique in his country, but I did not enjoy the American way of life. I learned later that in 1933, Morgan, at the suggestion of Cannon and Redfield, would have liked me to come to Pasadena, but Henri Fredericq and his friend the Rector J. Duesberg convinced him that it would be preferable for me to stay in Belgium. I have always felt a strong attachement to Wallonia and to Europe.

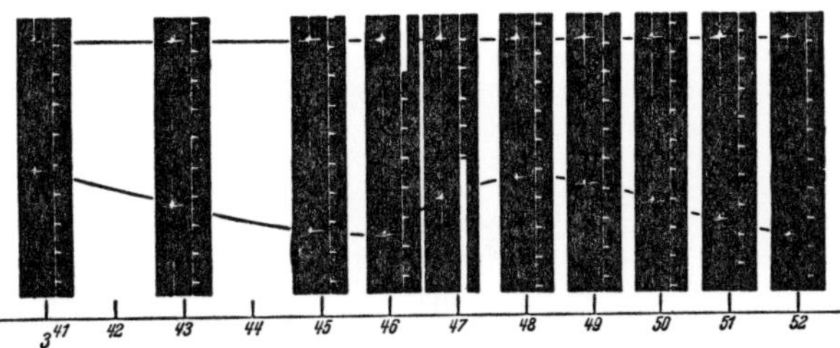

FIG. 9. — Parts taken from a continuous record of salivary secretion obtained from a denervated submaxillary gland of a cat and mounted vertically in such a way as to illustrate at once the changes in salivary secretion brought on by stimulation of the pilomotor nerves to the tail. Each vertical part shows on the left the interval between two drops and on the right, time intervals in 2 seconds. The rate of salivary secretion decreases (after injection of a small dose of pilocarpine) as seen from a comparison of the tracing taken at 3.41 and 3.45. During the minute 3.46 to 3.47, the sympathetic pilomotor fibres to the tail are stimulated for 30 seconds. The rate of salivary secretion increases slowly, reaching a maximum 2 minutes after beginning of stimulation (an increase of approximately 25 %), and then returns gradually to the pre-stimulation level. The experiment shows that 'sympathin', like adrenaline, stimulates salivary secretion in the denervated submaxillary gland of a cat.

(From Cannon and Bacq, *Am. J. Physiol.* 96, 403, 1931.)

Naturally, the research on 'sympathin'—the term Cannon proposed should be used until such time as the chemical structure of this substance was established—was actively pursued at Harvard and in Belgium, where, after a brief stay in Brussels, I moved to Liège with a grant from the 'Fonds National de la Recherche Scientifique'. I started from the highly probable hypothesis that sympathin was either adrenaline itself or a substance structurally closely related to it. Following the techniques used at Harvard, I demonstrated in the cat that sympathin augments glycaemia, inhibits intestinal motility, dilates the pupil, contracts the retractor penis, gives a chemical reaction characteristic of polyphenols, and was present in certain heart extracts from which it disappeared after sympathetic denervation.

Since my arrival in Liège I had been strongly attracted by the physical methods developed by Victor Henry. He welcomed me into his laboratory and we were able to show that a saline perfusate of the frog's heart taken during nerve stimulation absorbed intensely in the same ultraviolet region as the polyphenols, such as adrenaline. Furthermore, the absorption was strongly and irreversibly increased on exposure to an alkaline pH. This led to the logical conclusion that the substance in the perfusate had been oxidized to a quinone (Fig. 10). In order to appreciate the real importance of this observation, it must be remembered that at that time—40 years ago—there were no automatic Beckmann spectrometers. In order to obtain a U.V. absorption spectrum with Henry's method—the only one then available—it was necessary to record the bands of an iron-cadmium spark on a photographic plate during a period varying from 40 to 180 seconds, and then with the help of a magnifying glass to fix the intensity of those rays which were comparable to that emitted by a 10 second control. It took three or four hours to get a spectrum; we worked at night in order to have peace and quiet. We knew full well that these observations in aqueous solution would not allow us to characterize the specific absorption band or bands of sympathin, but they could show that a polyphenolic structure was extremely probable. In other words, I was unable to separate two closely related substances: adrenaline and noradrenaline, both highly likely candidates of sympathin. Among the general conclusions of my thesis (*Arch. Intern. Physiol.*, 1933) was the following (p. 242):

'Sympathin is surely an aromatic compound. It is probably either, like adrenaline, an amino-catechol derivative, or a mixture of amino-catechol derivatives in variable proportions.'

At Harvard, W. B. Cannon pursued his research with Arturo Rosenblueth, a Mexican physiologist, who arrived in 1930 just after we had completed the first series of experiments on sympathin.

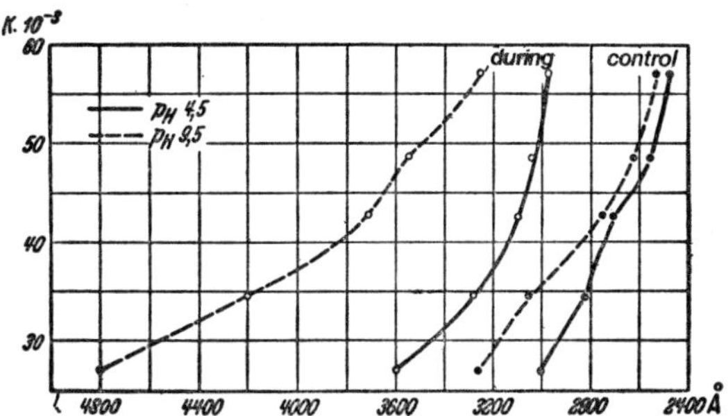

FIG. 10. — Absorption coefficients of two perfusates from a frog heart; one perfusate obtained before, the other during stimulation of the vago-sympathetic trunk. Solid lines: absorption in acid solution at pH 4.5. Interrupted lines: absorption in alkaline solution at pH 9.5. Abscissae wave length in Å. From this figure it is evident that stimulation of the cardiac nerves releases a substance which absorbs in the ultra violet range and that the absorption increases in alkaline solutions.

(From Bacq, *Arch. Intern. Physiol. 36*, 227, 1933.)

Rosenblueth introduced new and very useful techniques which were rapidly adopted everywhere in Europe: the nictitating membrane; cocaine sensitization—a phenomenon discovered by Loewi and Fröhlich in 1910; barbiturate anaesthesia—which eliminated the long preparation involved in the early experiments by Cannon and myself. Cannon and Rosenblueth perfected a fine technique which was very useful during the long period when the presence of minute quantities of adrenaline and noradrenaline could only be measured with biological tests. They introduced the simultaneous recording of the non-pregnant cat uterus (an inhibitory action of pure β-type) and of the nictitating membrane (an effect of a pure α-type) sensitized either by denervation or cocaine. It was with the use of this method and by a comparison of the effects of injected adrenaline with those caused by stimulation of hepatic sympathetic nerves that Cannon and Rosenblueth made the first and important observation that adrenaline did not reproduce the effects of sympathin in every respect, insofar as the inhibitory effects (β) on the uterus were clearly less pronounced than the contraction of the membrane.

It was unfortunate that the interpretation proposed by these authors in a famous publication in 1933 was not viable and that Cannon, Rosenblueth and their associates were led up a blind alley from which Rosenblueth stubbornly refused to extricate himself. The theory was as follows: The excitation of sympathetic nerve endings releases a mediator M, which is postulated to be adrenaline. This mediator M can combine with an excitatory substance E or an inhibitory substance I, localized in the receptors, to form two complexes ME or MI which would be the truly active substances. What would pass into the blood during sympathetic excitation would thus not be the substance M, but a mixture of ME and MI (sympathins E and I), in varying proportions. No physiologist competent in this field could accept this theory. I maintained my belief in Elliott's simple concept, and I was very sorry that I had to say in a report (1934) and in a review (1935) why Cannon and Rosenblueth's theory was unacceptable to me, and to propose another interpretation of the findings, which could not themselves be contested. Several hypotheses were plausible, but the one which I most favoured was the logical corollary of an analysis of the published observations. The fact demonstrated by Cannon and Rosenblueth was that 'Sympathin' released by the liver during excitation of nerves running along the hepatic artery was more active than adrenaline on the sensitized nictitating membrane (α effect) than on the non-pregnant uterus (β effect). In support of their concept of a purely inhibitory sympathin (Sympathin I), the authors quoted the following experiment: If the fibres leading to the duodenum, which are mixed with those which enter the liver were sectioned, the fibres running along the hepatic artery no longer had an inhibitory effect. A careful reading of the results and the text, however, convinced me that this was not true. By sectioning these fibres, Cannon and Rosenblueth reduced the amount of noradrenaline released by the liver to such an extent that the weak inhibitory effect of this amine would disappear completely. Out of this came my suggestion, in 1934, that Sympathin E could be none other than noradrenaline. I also had in mind another working hypothesis, which proved to be a blind alley and which I had to abandon after many rebuffs.

Following my return to Europe in 1932 I became acquainted with the highly significant paper by Barger and Dale on the relations between chemical structure and sympathomimetic action (*Journal of Physiology*, 1910, *51*, pp. 19-59). One fact stayed in my mind: DL-noradrenaline is more hypertensive than DL-adrenaline. One day while I was visiting Fourneau at the Institut Pasteur in Paris together with Tiffeneau, this fact was mentioned, and I can still hear Tiffeneau saying that he could not understand why 'nature' could not use a more active and simpler molecule than adrenaline.

It was not until much later in fact, after 1945, that on re-reading a paragraph of Barger and Dale's paper, I found, beautifully expressed, the suggestion that if Elliott was correct and there is a substance liberated by the sympathetic nerve endings, noradrenaline would be a better candidate than adrenaline. Here is the commentary that Sir Henry published in 1953 regarding this work (pp. 97 and 98 of *Adventures in Physiology*, Pergamon Press):

'In view, however, of the recent spate of discovery concerning noradrenaline, showing it to be, in many animals, the principal transmitter of effects from sympathetic nerve endings, it may be of greater interest to observe that I came near to that as a possibility even then, without actually recognizing it. I must take all responsibility for the failure; Barger did not even supply this material, for 'nor' adrenaline was already obtainable from the Höchst Dye Company, under the name 'arterenol'. I was chiefly concerned, however, with the difficulty created, for Elliott's theory of adrenaline as the sympathetic transmitter, and for certain rather fantastic corollaries which others had attached to it, by the fact that nor-adrenaline seemed to be much more accurately sympathomimetic. Doubtless I ought to have seen that nor-adrenaline might be the main transmitter that Elliott's theory might be right in principle and faulty in this detail. If I had had so much insight I might even then have stimulated my chemical colleagues to look for nor-adrenaline in the body; but they would almost certainly have failed to find it with the methods which where then available. It is easy, of course, to be wise in the light of facts recently discovered; lacking them I failed to jump to the truth, and I can hardly claim credit for having crawled so near and then stopped short of it. But if I had taken the additional step, even in hypothesis, much trouble might, perhaps, have been saved in after years for my late friend Walter B. Cannon! For the oberved differences between the effects of adrenaline and those of 'Sympathin', as liberated by stimulation of sympathetic nerves and especially of the hepatic nerves, which led Cannon and his associates to put forward the elaborate theory of the two complex sympathins, E and I, were practically the same as those between adrenaline and nor-adrenaline, with which I was here concerned. It should be noted indeed, that one who had collaborated with Professor Cannon at one stage, Professor Z. M. Bacq, actually suggested in 1934 that 'Sympathin E' might be nor-adrenaline (*Ann. de Physiol.*, 1934, *10*, 480); but, in the absence of direct evidence then, and for many years afterwards, of the natural occurrence of nor-adrenaline in the animal body, this suggestion attracted less attention than it deserved!'

Is it possible, with the aid of the published material and with the knowledge that I have of the personalities of W. B. Cannon and of Rosenblueth, to explain the origins of this erroneous concept of the 'two sympathins'? I believe so, and I will try to do it. Arturo Rosenblueth was an extremely intelligent physiologist, far more lively than Cannon's American associates. A good and hard working researcher, he also had a speculative mind and was in the highest degree interested in the mathematical expression of biological phenomena. He knew Wiener and launched himself into cybernetics. He might also have been impressed by the publications of Crozier who at that time worked at the Faculty of Sciences at Harvard, and had reduced animal behaviour—no matter what the animal—to an equation. Neither W. B. Cannon nor Rosenblueth had much grounding in basic pharmacology, they were not ready to convert themselves to the idea that the theory would require a reorganization of research, including a significant participation by biochemists.

To my mind, there is no doubt of Rosenblueth's personal responsibility in this unfortunate story of the two sympathins; he persistently defended the hypothesis long after the death of W. B. Cannon in 1945, although the evidence in favour of nor-adrenaline was so strong that no-one bothered to discuss the contents of his book published in 1950 (*The Transmission of Nerve Impulses at Neuro-Effector Junctions and Peripheral Synapses*) in which he indulged in a futile exercise.

Rosenblueth's interpretation led to a dead end. It was based on a certain notion of the 'receptor', the way Langley had formulated it, but its meaning at that time was vague compared to what it is today, though it captured the imagination of scientists. If one assumes that these substances E and I exist, but cannot put forward precise ideas about their possible chemical structures, the whole idea becomes useless as a working hypothesis. It prevents any progress, any design of new experiments. This is what happened. All the talk about the two sympathins consisted of useless theoretical discussions. It is unfortunate that it is always the work of W. B. Cannon and Rosenblueth that is quoted and torn to pieces when the subject of 'sympathin' is mentioned in the literature or in the classical textbooks and never the clear and faultless experiments Cannon and myself had published two years earlier. Now after forty years it seems as if W. B. Cannon's contribution to the problem of chemical transmission is dismissed as an error. How much more generous and nearer to the truth is the iudgement of Sir Henry Dale (see page 43).

And again there arises the same question: Why had we to wait years after the war to establish, thanks to U.S. von Euler, that noradrenaline was the principal if not the only transmitter of adrenergic neurones? I was not the only one who thought of this solution. In the U.S.A., several scientists had published observations in favour of the sympathin E = noradrenaline hypothesis (Stehle and Elsworth, 1937; Melville, 1937; Greer *et al.* 1938).

I did not feel easy at the thought of having to prove experimentally that my suggestion was right. I knew that my great teacher detested someone stepping on his toes. In his book *The Way of an Investigator* * he had expressed this sentiment clearly. At that time one was scrupulous, and there was 'fair play' in the world of biology. Morals have since degenerated. At present the competition is so intense that certain colleagues who are asked by the editor of a Journal to give an opinion about a manuscript would not hesitate to utilize the advance information in an unbelievable way, sometimes by using more perfected methods than those of the other laboratory. With a bit of luck and the editor's complicity, the secondary work may be published before that which inspired it. More than once during the last ten years I have been told the story of a participant of a Congress or Symposium who, after his return, decides to change completely the direction of the research in his laboratory after having heard an inspired but imprudent colleague propose a

---

* Here is the exact quotation: p. 41 'The investigator should be generous toward his fellow scientists. Generosity may be expressed in one respect by not rushing into another's field the moment it is opened. The successful explorer who has made startling progress in a novel enterprise should be given an opportunity to expand his discoveries without being crowded by newcomers.' and p. 42 'It is hardly necessary to mention a humble attitude as a qualification of the man of science. Even though it may not be an essential qualification, it is a highly desirable one.' Alas, the superiority complex which has invaded certain colleagues has supplanted such desirable generosity. However, a certain modesty is still part of the character and general attitude of the true research worker.

new, logical and exciting approach which he himself had not thought of. To give up without remorse a hypothesis, the usefulness of which is exhausted, is all right. But the way in which it is done is not always impeccable. When an idea has been developed by a research worker, with limited resources and a small staff, a European for instance, he'll be outclassed and surpassed in less than a year and the original contribution which was the beginning of an undoubted advance, may be quickly forgotten and not even mentioned in the references given in papers and reviews on this subject.

I do not hesitate to say that this unscrupulous competition has often discouraged and disgusted honest researchers to whom one should pay homage and that it explains why the field of chemical mediators after an essentially European beginning has become in five years a sort of inner circle, where the large American schools play the main role. One can foresee the arrival in force of the Japanese as in other fields of biochemistry.

But let us return to our subject. Before 1940, the chemical and physico-chemical methods were in their infancy and the scientific mind was in no way prepared to accept the idea that noradrenaline was the principal transmitter of sympathetic effects. In 1936, Loewi observed that 'Sympathicusstoff' extracted from frog heart gave the specific fluorescence described by Gaddum and Schild (*Pflügers Arch. ges. Physiol., 237*, 504). This remains true since Amphibians tend to N-methylate their amines. The case of the bufotenine in the toad is a typical example.* In Amphibians noradrenaline is not, as it is in mammals, the principal, if not exclusive transmitter. In 1939 (*J. Physiol., 96*, 385), Gaddum and Kwiatkowski made a very careful comparison between pure adrenaline and the 'sympathin' released from the isolated rabbit ear on nerve stimulation and found that neither

---

* The catecholamine extracted from the cutaneous glands of the toad is without question adrenaline. The bufotenine results from N-dimethylation of 5-hydroxytryptamine.

the chemical tests nor the two biological ones they used showed any differences whatsoever.

An amusing story shows how manifestly unready the minds were. We now know that for about 20 years the samples of natural L-adrenaline (extracted and crystallized from bovine adrenals) which were kept in many countries as standards for the Pharmacopoea, were contaminated with L-noradrenaline. But this only became clear with the development of methods which allowed separation of these two amines. In 1932 (*J. Physiol.*, *76*, 181), A. Szent-Györgyi (at that time still in Szeged, Hungary) published together with his co-workers a paper entitled 'The Function of the Adrenal Medulla' which begins as follows:

'We have reported in a previous note, that under certain conditions extracts of the adrenal medulla show an activity which cannot be explained by the adrenaline present. This activity was apparently due to a substance similar to adrenaline, but more potent than the latter. Till a chemical name can be substituted we propose to call this substance 'Novadrenine'.'

If one prepares an extract of freshly collected bovine adrenal glands according to certain techniques, the activity of this extract on the blood pressure of a decapitated cat is very much stronger (up to 10 times) than that of an adrenaline solution (it is not said whether it is L-adrenaline and uncontaminated) which gives the same intensity of colour by oxidation of the diphenolic functions (red quinone formation). This reaction should have given approximately the same colouration for either adrenaline or noradrenaline. At the same colouration the extract is also far more active on the isolated rabbit intestine than is adrenaline.

In the following year, U. S. von Euler published in the same *Journal of Physiology* (1933, *78*, p. 462) 'On the presence of Novadrenine in Suprarenal Extracts' in which the author repeats the experiments of the Hungarian physiologist, but using a better and more quantitative chemical method. The conclusion is the following:

'No significant difference was observed between the activity as found by biological tests (rabbit blood-pressure) and that found colorimetrically (using the method of Euler).' At the end of the paper, the following signed note is appended in small print: 'This paper was kindly submitted by Dr U.S. von Euler for my criticism. Von Euler is able to give an adequate explanation of the experiments of my collaborators and myself, without the necessity of supposing the existence of a more active form of adrenaline (novadrenine). I accept his view, by which the supposition of such a compound becomes superfluous.—A. Szent-Györgyi.'

On closer inspection, Szent-Györgyi's novadrenine was not noradrenaline, whose proportion in bovine adrenal extracts could not explain the observed differences. Even pure noradrenaline could not give such a difference. However, von Euler had the opportunity to see (just as Schild did a little later) that there was something other than adrenaline in the adrenal extracts.

The merit of having patiently accumulated the evidence in favour of noradrenaline is without question that of U. S. von Euler. The award of the Nobel Prize to him in 1970 was unanimously approved by his colleagues. The Swedish school is the only one, apart from the British, to have successfully resisted the North American invasion after 1950. The Swedish scientists have developed new chemical and cytological techniques. The beauty, the precision, the sensitivity and the usefulness, especially in the field of the central nervous system, of the histochemical demonstration of catecholamines and of 5-hydroxytryptamine by a fluorescent reaction, made possible magnificent progress in neurology and biochemical pharmacology. One must pay tribute to Falk, Hillarp and their associates for this exceptional contribution to the development of our knowledge in the field of the central nervous system which is so complex and technically difficult to tackle. These new conquests in fundamental science are basic for the interpretation of the effects of many drugs—classical and modern—which modify man's behaviour and which are at the present time often abused, the so-called psychotropic drugs: tranquillizers, neuroleptics, stimulants, etc. The list is long, and new substances are added every year. As in all human enterprise, the benefits are not without gross inconveniences. It seems, for example, unavoidable that the abuse of tranquillizers may produce a change in personality, a decrease of mental activity and other psychological changes.

The standardization of human behaviour, the profusion of contradictory information, the pollution of minds as well as of Nature—at least of what remains—cannot fail to disturb even the most optimistic among us.

6

The progress
of biochemical research

As soon as it was generally accepted that the effects of nervous excitation were caused by the release of chemical substances, the physiologists found themselves deprived of a vast field which passed into the hands of biochemists and pharmacologists. The biophysicists who, like B. Katz, have added the electron microscope to their arsenal, remain at the forefront of progress. The presynaptic structures (i.e., the nerve ending) and the post-synaptic structures, (i.e., the receptors localized on the membrane of the effector cell), have been studied in the minutest detail. These receptors are the structures which determine the end result—excitatory or inhibitory—of a nerve impulse.

The function of a neurone and its axon is to synthesize, store and release a specific substance. Biochemically multifunctional neurones releasing, for example, both acetylcholine and noradrenaline, have not as yet been discovered. This possibility, however, still exists.

6.1. ACETYLCHOLINE SYNTHESIS AND STUDIES OF CHOLINESTERASES

It seemed clear that the biochemists would have to elucidate the mechanism of acetylcholine synthesis and to study in detail the cholinesterases and their inhibitors. It was also necessary to explain how acetylcholine could be stored in cholinergic nerve endings and be protected from the action of the cholinesterases. Chemical logic suggested that choline—a common molecule in all living beings—was acetylated enzymatically. Physiological proof was provided by G. L. Brown and MacIntosh: if a sympathetic ganglion is perfused for a long period with a saline solution which contains no glucose (energy) and no choline (precursor) the transmission of excitation from the preganglionic fibres fails. To re-establish transmission, it is necessary only to add choline and glucose to the perfusing fluid.

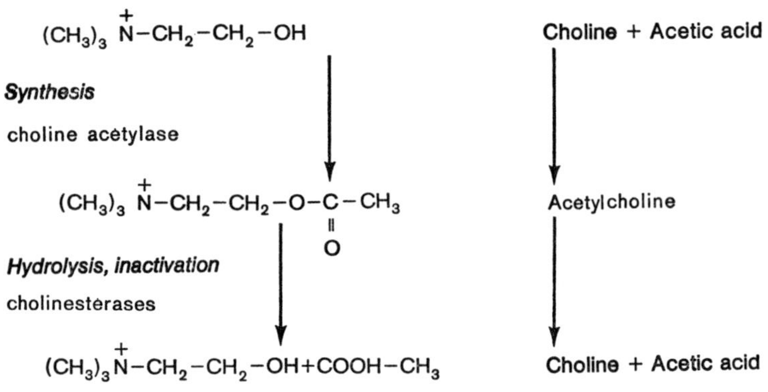

In 1936-37, Quastel and his collaborators, as well as Stedman and Stedman, demonstrated the presence in the nervous system of an enzymatic mechanism for acetylcholine synthesis. Feldberg and T. Mann (1944) using an extract of brain acetone powder confirmed this, and in 1946 mentioned the existence of an activator which was later isolated and identified by Lipmann. This discovery of the now famous coenzyme A, necessary for the transfer of acetyl, propyl, benzyl, etc. radicals, won for its author the Nobel Prize. In fact, the acetylation of choline is only one of the manyfold examples of chemical reactions catalyzed by coenzyme A; all these exchanges take place at the region of the terminal SH group of the molecule.

Nachmansohn * and his co-workers demonstrated the importance of ATP (adenosine 5'-triphosphate) in acetylcholine synthesis; in fact, they might have identified coenzyme A before Lipmann. They were more successful, however, in their work on the cholinesterases. At the beginning of the 1939-45 war, the Nazis had manufactured large quantities of two substances, Sarin and Tabun, whose physical and chemical properties qualified them as potent war poisons. The British (Adrian and Feldberg, playing their part in this discovery) knew that these organophosphates, toxic substances, inhibited cholinesterases in an irreversible manner. The merit of Nachmansohn's team (Wilson, Leuzinger, Carlin...) was to study this phenomenon at the molecular level and to develop logically (but long after the end of the war) the preparation of true antidotes, i.e., substances capable of displacing the anticholinesterases from their combination with the enzymes and regenerating enzyme activity.** Nachmansohn, quite unsuccessfully, defended the idea that acetylcholine is involved not only at the nerve ending but also in impulse propagation along nerve fibres, whether or not they are cholinergic. At the present time, however, the biophysicists have described this phenomenon as one involving rapid ionic exchanges at the level of the nerve membrane.

The complex biochemical properties of cholinesterases were soon realized. A true or specific acetylcholinesterase was isolated; it is the most important one from a physiological point of view, it is absent from plasma and differs from a non-specific cholinesterase, especially in its sensitivity to various inhibitors. Failure to distinguish between these two different enzymes would leave one open to errors of interpretation; certain papers unfavourable to the theory of cholinergic transmission have their origin in such errors.

---

* German biochemist: emigrated to the U.S. (at Columbia University in New York) after a brief stay in France and in Great Britain.

** Sir Rudolph Peters during the war had succeeded in the first attempt of this nature: to regenerate by means of a dithiol, the famous BAL (British Anti-Lewisite), the functions of thiol enzymes blocked by Lewisite arsenoxide.

6.2. CATECHOLAMINE SYNTHESIS

The synthesis of catecholamines is far more complex than that of acetylcholine. This was known for a long time, but it was necessary to wait until the 1950's and the extensive use of radio-isotopic labelling to learn in detail the processes which regulate the synthesis of dopamine, noradrenaline and adrenaline from two amino acids present in all living beings. The North American biochemists and pharmacologists had these labelled substances at their disposal well before the Europeans; and this is one of the reasons for the recent rapid advances they made in this field, in which European biochemists, Blaschko for example at Oxford, had made important contributions.

This story is, therefore, more recent than that of the great physiological discoveries of the pre-war period. It is still far from complete, many of the steps and above all the sequence of events remain the subject of heated discussions among specialists. Our knowledge can be summarized as follows (Fig. 11): Phenylalanine is the most common precursor, two successive hydroxylations transform it into dihydroxyphenylalanine or DOPA.

The enzyme tyrosine hydroxylase which catalyzes the second hydroxylation is present in only low activity in adrenergic cells. The conversion of tyrosine to DOPA is thus the rate-limiting step in the synthesis. DOPA is then decarboxylated to dopamine by an enzyme which is found in many tissues; dopamine is hydroxylated in the beta-position of the lateral chain by a highly specific enzyme dopamine-β-hydroxylase. After four steps noradrenaline is thus obtained from phenylalanine. To obtain adrenaline one only needs a methyl group to be taken from the universal activated donor, S-adenosylmethionine and to be transferred on to the amino nitrogen by an N-methylase enzyme.*

---

* Recent experiments seem to indicate that the methyl donor for the N-methylation of dopamine in the brain is 5-methyltetrahydrofolic acid.

The processes of hydroxylation, methylation and demethylation are relatively complex; they differ from one tissue to another and from one animal species to another. The synthesis can stop with dopamine; there is rarely a second methylation on the

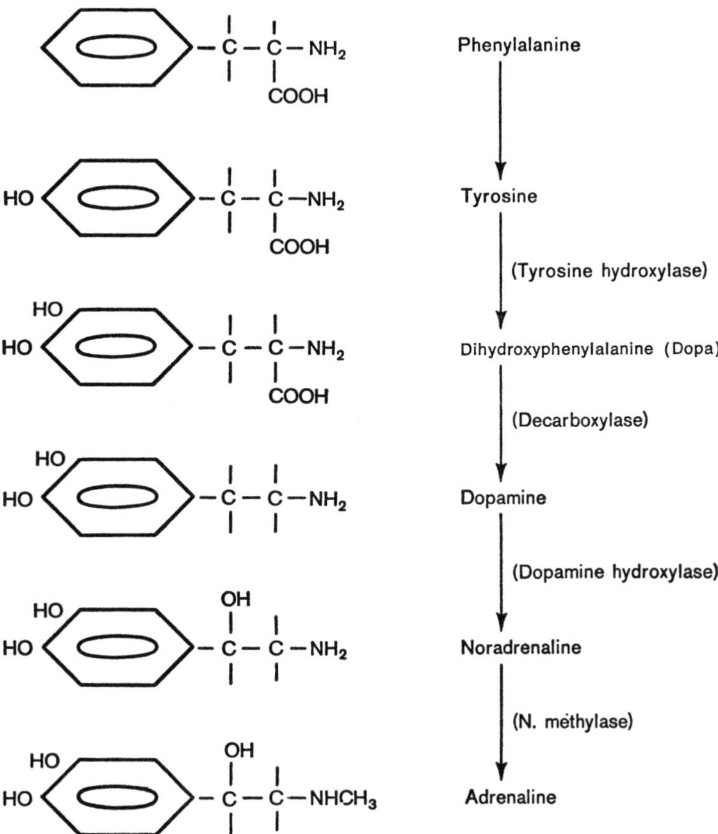

FIG. 11. — Catecholamine synthesis in Vertebrates.

nitrogen. In the Cephalopods, tyrosine is decarboxylated to p-tyramine which can be β-hydroxylated to form octopamine. The strategic localization of the enzymes within the cell, the mechanism protecting the various mono- and diphenolic substances from enzymatic inactivation, the storage and release of these amines are still the objects of study for many research teams throughout the world. As with acetylcholine, it is thanks to the remarkable physical properties of specific storage vesicles that all these processes can occur. Their granules are visible with the electron microscope, a wonderful instrument which has at last enabled biochemists, biophysicists and pharmacologists to find a common frontier with the histologists. One now has very precise ideas of biochemical anatomy and a reconciliation is in progress between anatomists and biochemists whose language, concepts and mentality were completely different 20 years ago.

7

The opposition to the theory of chemical transmission

Study of the criticisms of the chemical transmission theory from 1921 (the date of Loewi's first publication) until 1945 (the date of J. C. Eccles' conversion) is perhaps even more interesting than an account of the accumulation of supporting evidence. Opposing arguments of all kinds, both theoretical and experimental, were invoked and naturally it was O. Loewi who was most severely attacked. The reader will find the exact details and references in my review (1935) and in Lembeck and Giere's book.

Before 1926, several authors had difficulty in reproducing Loewi's milestone experiment of 1921. For five years Leon Asher of Basel engaged Loewi in a lively polemic, and when I met Asher after the publication of my review on the chemical transmission of nervous impulses in the *Ergebnisse der Physiologie und experimentellen Pharmakologie* (1935) * he held that Loewi had tried to defend an exact theory with mediocre experiments, and that the discovery of the properties of eserine was little more than a stroke of luck. Asher claimed to have discussed in 1907 and 1910 with his co-workers the hypothesis that two different substances

---

* Asher liked it. Unfortunately, it was published in French and even at that time this was a severe handicap. It would have been read more widely had I decided to write it in English.

might be released from the endings of the excitatory and inhibitory cardiac nerves. In 1917 Asher carried out experiments in his laboratory, using a technique similar to Loewi's; however, the results were negative. Asher was perhaps an unlucky forerunner. It is always unpleasant to watch a colleague demonstrate successfully an idea which one could not prove oneself. Loewi's demonstration, which he was asked to perform at the 1926 International Congress in Stockholm, did not convince all observers, but Loewi declared himself satisfied.

Attempts to obtain evidence for the chemical transmission of the effects of cardiac vagal stimulation in the dog, using various cross-circulation techniques, regularly failed. The experimenters, notably F. C. Heymans in Ghent and Tournade in Algiers, lacked neither skill nor experience. Without eserine the cholinesterases in the mammalian heart and blood are so active that no acetylcholine can escape from the tissues in which it is released. The experiments had to fail. One only needed to add physostigmine, as did Feldberg and Krayer, or to perfuse the heart with a saline solution, as did Rijlant, Foa and Vale, to get a positive result.

Numerous authors claimed that 'Vagusstoff' did not exist, and that Loewi's observations could be explained by an increased K^+ ion concentration in the saline solution during vagal stimulation. There is however a pharmacological argument which deals with this objection: atropine abolishes the actions of vagal stimulation and of 'Vagusstoff' but not that of K^+ ions. It was soon discovered (Loewi and Navratil, 1924) that atropine does not prevent the release of acetylcholine from cholinergic nerve endings. The French physiologists, Tournade, Hermann, Chabrol and above all, Malméjac—who at the time worked in Algiers—produced more restrained criticisms; they had been able to reproduce Loewi's observations in the frog, but not in the tortoise or the dog, using cross-circulation techniques. It took about ten years to dispel the notion that 'Vagusstoff' was specific for vagal effects and only localized in the heart. Loewi can hardly be criticized for naming the substance whose actions he had observed 'Vagusstoff'. What other term could one have used in 1921, before its chemical identity with acetylcholine and wide distribution in all organs had been established? Malméjac, in his painstaking thesis of 1928, argued that he had found an action resembling that of 'Vagusstoff' in extracts of tissues which contained no vagal fibres.

With the perspective of half a century, the progress and exact expression of our ideas seems startingly slow. It was only in 1933 that Dale proposed the use of the terms 'cholinergic' and 'adrenergic' in the autonomic nervous system, the first step towards a biochemical definition of nervous function whose consequences have already been pointed out.

Although from 1921 until 1931 Loewi bore the brunt of criticism, he was not alone. For example, Henri Fredericq published a paper in 1925 entitled 'Le mécanisme humoral des actions nerveuses' in the *Revue des Sciences de Paris*, and presented (1927) a report entitled '*La transmission humorale des excitations nerveuses*' * to a plenary session of the Société de Biologie in Paris. Using the chronaxy of cardiac muscle excitability as a test, he had regularly obtained positive results in 1925 and in 1926. The severity of criticism was such that Henri Fredericq did not actively continue his research in this field until my arrival in 1932. Louis Lapicque, for example, with whom H. Fredericq had worked in Paris, declared Loewi's theory to be 'unthinkable'. Henri Fredericq gave me every facility in order that I might complete my thesis in 1933. We subsequently worked together and published many papers and notes. No boss could have been more generous or more attentive to the career and happiness of his associates.

---

* These terms were in current use at the time: they translate very exactly the title used by O. Loewi in his series of publications: *Über humorale Übertragbarkeit der Herznervenwirkung*. Again, it took about twenty years to understand that humoral transport was only a phenomenon of secondary importance, although technically useful in the study of chemical transmission.

In 1933, the 'Association des Physiologistes', at the suggestion of André Mayer, chose the pharmacology of the autonomic nervous system as the main theme for its 1934 meeting at Nancy, and asked L. Lapicque and myself to give a review of the subject. Mayer continued to exert great influence in the Association which he had helped to create, and whose publications he had subsidized. One of its goals was to prevent a loss of contact between French physiologists and their Belgian and Swiss colleagues, many of whom often visited Great Britain and the United States. The task which he assigned to me was pleasant enough from a scientific point of view, but very difficult in that my co-reporter, L. Lapicque, was well known both for his ability in debate and for his strong opposition to the idea of chemical transmission. His discussion remarks at the 1933 Liège Congress of French speaking physiologists had impressed me deeply. I was only thirty-one and Louis Lapicque had long been the undisputed leader of neurophysiology at the Sorbonne. I foresaw a most harrowing and difficult confrontation.

Although I have never run away from a real argument, I have always tried to keep it from becoming personal. I knew Lapicque only through his work. At the end of 1933 we had a long meeting in his office in Paris, he shut his door to avoid any interruption. I will always have a moving memory of the debate which ensued and lasted for more than an hour. Happily the logic of my arguments converted Lapicque, and he announced to his somewhat surprised associates that I had convinced him of the validity of the chemical theory *as far as the autonomic nervous system* was concerned. At that time no-one thought of the possibility of chemical transmission at the motor nerve ending or from neurone to neurone. This visit to the Sorbonne was the origin of my work with A. M. Monnier and of a new and charming friendship. I was soon told that it had been my good fortune in the discussion with Lapicque that I was young and not of French nationality. Never, I was told, had a colleague of his age or a compatriot succeeded in influencing Lapicque in that way. I must admit that here again, fortune had smiled upon me.* When I saw L. Lapicque again in 1946, he was very happy to learn that a French speaking physiologist had shown in South America that European physiology was not crippled, as certain people would have believed it to be.

The most spectacular and useful criticisms undoubtedly came from J. C. Eccles, a brilliant physiologist, who from 1932-35 had worked with G. L. Brown. After emigrating to Australia and New Zealand, he returned regularly to Great Britain, and the controversy between his group and Dale's imposing school of thought much excited the members of the Physiological Society. The discussions never ended at Hampstead, and both experimental and theoretical arguments were prepared to counter the criticisms he raised. Eccles obstinately forced the partisans of the chemical transmission theory to defend their theory solidly, and thus to produce direct evidence in its favour. Such important physiologists as Lorente de Nó (*Amer. J. Physiol.*, 1938, *121*, 331) flocked under Eccles' banner. Lorente, perfusing a sympathetic ganglion with eserinized blood could only occasionally detect acetylcholine in

---

* On many occasions I had the pleasure of meeting Louis Lapicque in Paris, at his home at l'Arcouest, and again at Plymouth in 1936 under the following circumstances: To occupy my free time between two series of experiments in London, I conducted several experiments on Invertebrates at the Laboratory of Marine Biology at Plymouth. The now famous electrophysiologist Fessard also worked there. A. V. Hill was the natural patron of this laboratory, which has rendered such great services to English Physiology. With his permission, Fessard and I invited L. Lapicque to come to Plymouth in his famous two-masted yacht 'Axon'. Lapicque's answer: All right, when the weather is favourable. Demonstrating the yacht's capabilities and his navigational skill, one fine morning the *Axon* was sighted in the outer harbour. Louis Lapicque disembarked with Mrs Lapicque. Their crew consisted of one seaman and one cabin boy. A. V. Hill organized a reception at his lovely estate 'Seven Hills'. We arrived with the Lapicques. A. V. Hill greeted them very cordially and praised their courage for having crossed the Channel in the *Axon*. 'I did not believe that you would visit us in this fashion.' Lapicque replied without hesitation: 'My dear colleague, you never believe anything I say.' Everyone knew that the basic concepts of Lapicque and Hill had nothing at all in common. This anecdote was related by A. V. Hill.

the perfusate after stimulation of the pre-ganglionic fibres. Eccles was the one to whom all the non-believers made reference when they had run out of arguments, until his 'conversion' around 1945. This points to the importance of his 'conversion' which signalled the down-fall of the classical physical theory—a theory which Dale maliciously described as 'Ecclesian'—which held that eddy currents at the axonal endings directly excited the neuromuscular junction of striated muscle or influenced the membranes of the neurones at the synaptic level. Eccles used essentially physical methods, he recorded action potentials by *potential* measurements. One believes in what one sees with one's own techniques; chemical and pharmacological arguments could not shake his firm belief. Eccles had not been tormented, as had been Loewi, Feldberg and so many others by the need to provide a logical, plausible explanation of the inhibitory effects of excitation of the cardiac vagus, of muscarine and the choline esters, and of the ability of atropine to suppress all these effects. It seems that psychological barriers exist in the minds of scientists, which prevent certain facts or ideas from penetrating into daily thought and activity.*

The following sentence may be found in the last of three communications presented by Eccles and O'Connor, March 11, 1939, at the Physiological Society: '.... Thus we have been unable to find any unequivocal evidence that eserine action is related to the inhibition of cholinesterase with the resulting slower destruction of any acetylcholine liberated by nerve impulses at the neuro-

---

* There would be much to say about the weight a scientist accords to favourable or unfavourable arguments with respect to his basic working theory. It is always much easier to assess the value of an argument based on experimental techniques with which one is familiar. Some physiologists give little weight to arguments based on biochemistry or pharmacology. It is not clear if this is due to caution or to a slight superiority complex, which determined the basically neutral attitude of many of my able colleagues. Who, in retrospect, can deny the dominant role pharmacological observations have played in this history of chemical transmission? However, all disciplines intertwine and complement each other, physiology, pharmacology, biochemistry, electron microscopy. None of them can claim supremacy.

muscular junction.' (*J. Physiol. 95*, 37-38*P*). Eccles tried to refute the solid arguments provided by G. L. Brown, Dale, Feldberg and myself from 1935 to 1938, by invoking a direct excitatory action of eserine on the motor end plate, which he compared to that of veratrine on the nerve fibre. In 1937, however, Brown and I showed that the action of veratrine had nothing in common with that of eserine.

During the war Eccles published, in the *Journal of Physiology* (1943, *101*, 465, and 1944, *103*, 27), papers on the nature of synaptic transmission in the sympathetic ganglion. No mention is made of the experiments of Dale and his associates; the term 'acetylcholine' does not appear a single time; one is referred instead to the work of Katz, Kuffler, Lorente de Nó and to Eccles' own work. The introduction to the paper, submitted in October 1942, is quite typical: 'It has now been established that neuro-muscular transmission is mediated by the local negative potential which a nerve impulse sets up at the end plate region of a muscle fibre ... Similarly, it has been shown that synaptic transmission may be mediated by the local negative potential...' The analysis of synaptic potentials after curarization was alleged to confirm the basic theory. At the same time, Feldberg concluded from his study of acetylcholine synthesis in the sympathetic ganglion and cholinergic nerves: '...the synthesis of acetylcholine in sympa-thetic ganglia is a property of the pre-ganglionic endings and a necessary preliminary for normal and particularly sustained synaptic transmission.' (*J. Physiol.* 1943, *101*, 445).

In his work Feldberg ignored Eccles' arguments, and he did not even bother to refute them. The two groups knew one another, but there was a wall between them; each pursued his own research with his own basic concepts and methods. This ability to con-sciously exclude each other's views so rigidly was astounding, the more so in that the two papers appeared in the same issue of the *Journal of Physiology*. Perhaps the war, which geographically separated Great Britain from Australia, was in part responsible for this state of affairs.

In 1944, Eccles began to show some concern;* he was no longer as affirmative or as trenchant. His paper 'The nature of synaptic transmission in a sympathetic ganglion' (*J. Physiol.* 1944, *103*, 27) began as follows: 'It has been shown that synaptic transmission in the stellate ganglion is largely if not entirely, mediated by the local catelectronic potential which preganglionic impulses set up in ganglion cells...' '...in order to throw light on the role of acetylcholine...' This time he discussed the observations of the 'chemists'. He admitted that eserine considerably prolonged the synaptic potential obtained by preganglionic stimulation, and that this prolongation was undoubtedly due to acetylcholine. However, according to Eccles, the brief action potential was provoked by action currents. In fact, it was not until after 1945 that Eccles seemed less than satisfied. The observations he made using his own methods became less easily interpretable by his 'eddy current' theory. The electron microscope had demonstrated the existence of specific vesicles at the axon endings and the histochemists had demonstrated the exceptionally high concentration of cholinesterase in the postsynaptic membrane. The weight of evidence became daily more impressive. Eccles explains the details of his long and difficult 'conversion' process in his book *Facing Reality*.

It was on reading the works of Karl Popper, and on visiting the Austrian philosopher exiled in New Zealand in 1943-44, that Eccles convinced himself of the fact that the scientist grows in stature by recognizing what must be called in the present case, an error of interpretation.**

---

* In a 'confession' published much earlier than *Facing Reality*, Eccles declared: 'Around 1945, it became clear to me (although I did not want to admit this in public) that my hypothesis was in bad shape'. *Nerve as a Tissue*, Kodahl and Issekutz, Edit., Harper and Row, New York, 1966, p. 446.

** I thank my friend P. Devaux who translated into French, Popper's work *The Logic of Scientific Discovery* and explained the reactions of the Eccles-Popper pair to me. Popper quotes Eccles. Eccles quotes Popper at length in his book, *Facing Reality*, in other publications and in his

Politicians believe that they lose face and are discredited if they admit to having defended an unworthy cause. Here the stakes were formidable. Eccles had to abandon the basic concept which he had held until then, in order to adopt the diametrically opposed point of view of his opponents. Probably those four years of hesitation and reflection were beneficial, as Eccles subsequently, in the role of the ardent neophyte, resolutely placed himself in the vanguard of the research, much to the great joy of Sir Henry. His audacities stunned the old guard. Eccles had changed mounts, the rider was as tempestous as ever, but the horse was much better. A Nobel Prize crowned his achievements in 1963. A fine lesson: On the one hand, one must fearlessly say what one thinks

---

lectures notably during the one he gave in Brussels, May, 1973, at the bicentennial of the Academie Royale de Belgique. Quotation by J. C. Eccles (Nobel Lecture): 'I can now rejoice even in the falsification of a cherished theory, because even this is a scientific success.'

Falsification is a key word in the Popperian vocabulary; it has lost its pejorative connotation (at least in French in which falsification is a word currently used in the meaning of adulteration of food stuffs or of faking an experimental protocol). For instance, according to Popper and Eccles, one must look not for a confirmation of the working hypothesis, but rather to showing its weaknesses and to demolish it. When an incompatible fact has been observed, something new and interesting has been found. This, after all, is the attitude one takes, or which is taken by the majority of physiologists worthy of the name, who unconsciously seek 'falsification', as M. Jourdain did with prose.

One of Popper's methodological principles is that it is through 'falsification' of our preconceived ideas that we enter into contact with reality. According to P. Devaux, Popper is in reaction against the neopositivists, who are too fascinated by the unattainable original fact and by their nominalistic conception of reality. In a recent book, (*The Scientific Approach*, Academic Press, London, New York, 1973), J. T. Davies considers, as does P. Devaux that Popper's contribution to the philosophy of science is most important.

However, the fact remains that Popper's philosophy did not influence any of the other scientists involved in the research discussed here. Perhaps Eccles represents a special case, where a philosophical idea came to the aid of a scientist in making a volte-face. I never heard Cannon, Dale, Gaddum, Feldberg, Brown or any of my other friends allude to the need for their research to lean on a philosophical theory. Was this perhaps on their part, the mark of extreme discretion?

and on the other, declare openly if one has changed one's opinion. One would have to be obsessed by an immense vanity to assert that one has never done or written anything wrong, or made a mistake. One's errors aid others in finding the right way. One can err brilliantly, but one can also, as did Paul Claudel, turn one's conversion into a spectacular theatrical event.

At the end of his career, Sir Henry took pleasure in republishing in one volume his principal publications between 1906-38 (*Adventures in Physiology*, Pergamon Press, 1953). One can only pay tribute to the experimental precision, the farsightedness, the astonishing sagacity and flair of this exceptional man who had no need of any conversion and who was able to pass serenely the long years of an active retirement. The interest of this volume, however, is centred upon notes which he wrote in 1952 for these reprinted papers, in which he modestly attempted to explain how, in certain cases, he overlooked certain important facts—for example, the presence of ergometrine in ergot extracts (see also p. 43). The privileged few who had the opportunity to explore to the fullest the field of autopharmacology during the years 1930-45 are the only ones actually able to assess the greatness of Sir Henry's personality. Without any doubt, no-one influenced the evolution of physio-pharmacological sciences more profoundly in the first half of our century. Feldberg wrote a biography of Dale for the *Biographical Memoirs of the Fellows of the Royal Society* (Vol. 16, Nov. 1970, 77-174) which is a model of style, precise, complete, warm and admirably modest. No other genius has received a comparable tribute from one of his co-workers.

C. Heymans refused to accept the concept of a substance interposed between the nerve ending and the effector cell, at least until 1949, but for him it was a secondary issue. He was a physiologist and pharmacologist of the classical mould, a skilful experimenter and tireless worker, head of an energetic group. His work as it grew became of great importance and earned him the Nobel Prize in 1939. His research was mainly on respiratory and cardiovascular reflexes and the role of the carotid sinus and carotid body. He was not directly interested in the chemical transmission theory, as his experiments were easy to interpret whatever the nature of the theories attempting to explain transfer of information from nerve to muscle. On my return from the United States in 1931, an active collaboration was established between the physiologists of Liège and Ghent. I brought new methods to Europe, methods useful for the study of the motor component of the vascular reflex arcs—notably total sympathectomy. I carried out, often alone, sometimes with a close friend, Lucien Brouha who went to the United States during the war, experiments on the physiology and pharmacology of the autonomic nervous system, all of which produced evidence in favour of the chemical transmission theory. I knew that C. Heymans did not like them, and a certain respectful friendship prevented me from discussing them openly with him. The hazards of the war were such that, in 1939, I had to work on problems posed by chemical warfare, and after the armistice I was occupied by the Ministry of the Interior with problems concerning the protection of the civilian population. The lacrymogens and vesicants were my favourites. The radiomimetic effects of mustard gas led me towards ionizing radiations, and to the discovery of the phenomenon of chemical protection against these physical agents, and in 1952, with the collaboration of P. Alexander, to a re-orientation of radiobiology. The scientific branch of the Belgian armed forces engaged C. Heymans to study the effects of the nerve gases manufactured in large quantities by the Nazis, but fortunately never employed. These organophosphates had already been studied in depth during the war in Great Britain as well as in the United States; their powerful inhibitory action on cholinesterases was known, and, thanks to Nachmansohn and his co-workers some idea of their biochemical pharmacology at the molecular level had been established and accepted by the majority of specialists.

C. Heymans revealed his opposition to our general concept by publishing, from 1946 to 1949, a series of notes and papers in which he claimed to have demonstrated that the pronounced inhibition of cholinesterases by these phosphate esters was not the cause of the observed toxic effects. His main argument was that a total inhibition of esterases could be demonstrated in the blood and in heart extracts after injection of DFP (diisopropylfluoro-phosphonate) without causing any adverse physiological effects.* A logical, although as yet unproved, reply can be given to this argument: The esteratic activity of heart homogenates is not neces-sarily an exact measure of the conditions existing *in situ* where the enzymes are highly localized. A large margin of safety exists; an excess of enzyme molecules may require a very extensive inhib-ition before it leads to the appearance of physiological disorders. As C. Heymans had presented some of these observations at the Académie de Médecine de Belgique to which I belonged since 1946, I could not ignore them. My colleagues knew of my attach-ment to biochemical pharmacology. Since 1930 my research activ-ity had largely centred around this concept. Given my character, a discussion of these fundamental problems became inevitable. I brought it about by presenting to the Academy, together with my friend and colleague R. Weekers (Professor of Ophthalmology at the University of Liège), the results of a long study of the funda-mental pharmacology of the eye, in which we demonstrated that only the chemical transmission theory allowed a logical inter-pretation of the results. Heymans reacted vigorously. The inter-ested reader can find the details of this heated and prolonged debate in the *Bulletin de l'Académie de Médecine de Belgique* (6th series, *14*, 1949; pp. 76, 183, 299, 379, 383, 484, 489).

In re-reading these papers, now almost thirty years old, I am still astonished to see how many distinguished minds of the time remained obstinately closed to the biochemical interpretation of the effects of toxins and hormones, and to the basic concept of

---

* QUOTATION: 'We have observed that cholinergic symptoms can disappear in the animal intoxicated by DFP, while the activity of the two cholinesterases remains minimal or null.' (*Bull. Acad. Med. Belg.*, 1949, *14*, p. 77.)

chemical transmission. This idea, which to me seemed self-evident and not in need of any further defence, was still a sort of heresy to certain minds, including many eminent ones. My position was quite delicate. Academies rarely tolerate novel ideas. But had the Nobel Prize of 1936 not been given to Loewi and Dale thirteen years prior to this discussion? Did I not have the right to present my views in an objective manner, without being accused of impudence? Only Heymans, Weekers and myself were involved in this long discussion, which had taken on the appearance of a lone crusade. I deeply resented the impression of being repudiated, and accused behind the scenes of immodesty by the members of this respectable Academy. Heymans was still invoking Eccles' opinions when all other informed physiologists knew that he had undergone 'conversion'. The last bastion of resistance had fallen.

After 1950, C. Heymans began little by little, to speak of cholinergic and adrenergic nerves, like everyone else. I met him regularly at meetings of Belgian societies and in various pharmacological committees. We remained close friends to the profound astonishment of those traditionalists who thought us to be estranged. Why did Heymans and Eccles object for such a long time? In my opinion it could be because they did not have the biochemical background that Dale had, and so, quite naturally, biochemical and pharmacological arguments made little impression on them. Their techniques were those of classical physiology; these like Eccles' electrophysiological methods impeded their vision and created a barrier. Like Eccles, Heymans was not accustomed to thinking in biochemical terms; the logic and intrinsic beauty of this science was foreign to him.

I should add that it was my good fortune to meet, at the 'Institut L. Fredericq', an eminent biochemist, M. Florkin, who had been my friend since we first met at Harvard in 1929. A. Mayer, with whom I had spent a year, was even then, one of those physiologists thinking in biochemical terms. The friendship of V. Henry, the presence in Liège in 1939 of Szent-Györgyi, the meeting in

1945 with Sir Rudolph Peters at the Physiological Society exposed me to a continuous recycling process. I was forced to see the value of biochemical arguments. But I also realized that with my lack of interest in electrophysiological techniques, I risked overlooking important facts which these methods might unearth. I had to watch myself. Have I always reacted in due time in the right direction?

8

Some blind alleys

In the course of research on the chemical nature of sympathin, I was responsible for two series of experiments which led to an impasse, although they yielded useful information. The first arose from the following experimental fact: Rosenblueth had observed that cocaine, which he had used to sensitize the nictitating membrane, also augmented the response of the smooth muscle to sympathetic nerve stimulation. However, he had not undertaken any quantitative analysis of this phenomenon, even though this muscle lent itself perfectly to this type of research.

In studying in more detail this phenomenon of cocaine sensitization I was struck by the fact that cocaine enhanced the action of epinine (a molecule closely related to adrenaline, lacking a β-hydroxyl in the side chain), 4-5 times more than that of adrenaline. The comparison on the same membrane of the sensitization by cocaine to nerve stimulation on the one hand and to that of a sympathomimetic amine on the other was an indirect method of clarifying the chemical structure of sympathin. The first observations of H. Fredericq and myself showed that in order to obtain consistent results, it was necessary to eliminate a small cholinergic component by treating the cat with atropine. Our findings have often been quoted and interpreted in various ways. As was postulated by our working hypothesis the curves representing the response to increasing doses of L-adrenaline and to maximal stimuli at a frequency of 200 per minute were displaced to the same extent towards the left by cocaine, while in the comparison of epinine with nerve stimulation cocaine showed a greater effect on the epinine curve than on the effects of nerve stimulation (Fig. 12).

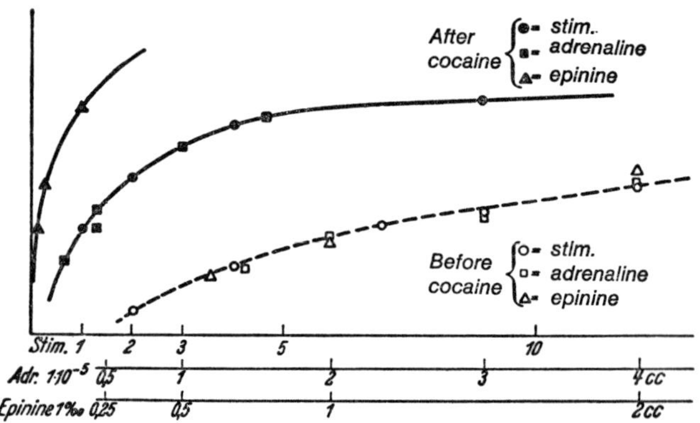

FIG. 12. — Cat under dial anaesthesia. Effect of cocaine on the responses of the nictitating membrane to L-adrenaline, stimulation of the cervical sympathetic and epinine. Ordinates: isotonic contractions of the membrane. Abscissae: number of supramaximal stimuli (condenser discharges 20 volts, 11 μF, frequency 200/sec). Beneath doses of L-adrenaline and epinine injected intravenously. The results are arranged so that the values obtained for nerve stimulation, adrenaline and epinine before cocaine fall on the same (interrupted) curve. It is clear that the cocaine sensitization is the same for adrenaline and nerve stimulation, i.e. the adrenergic transmitter. In contrast, the sensitization is much stronger for epinine which differs from adrenaline only by the absence of a secondary alcohol in the side chain.

(From Bacq & Fredericq, *Arch. Int. Physiol.* 40, 304, 1935.)

Conclusion: Sympathin is not identical to epinine. Fredericq and I also studied D-adrenaline and racemic noradrenaline, which we found in the Institute stores. At the time, it seemed to us that L-adrenaline best fitted the requirements of the working hypothesis.

Should we have seen at that time that noradrenaline reproduced the effect of nerve stimulation more faithfully than adrenaline? Had we worked badly? No. At the time (1934) the two optical isomers of noradrenaline had not been separated and our natural L-adrenaline was probably contaminated with noradrenaline. Our indirect method allowed us to exclude epinine, adrenalone, and non-diphenolic amines, but it did not allow any distinction between adrenaline and noradrenaline. There was a rebirth of interest in epinine when it was shown that dopamine could be N-methylated before being β-hydroxylated. One of the actions of cocaine is to inhibit the active biochemical uptake which replenishes the storage vesicles with free molecules reabsorbed from the interstitial fluid in the synaptic gap. It is not yet known whether epinine is more actively reabsorbed than the other amines, particularly noradrenaline.

The other blind alley is the saddest story of my career. It is not yet explicable. Starting from the hypothesis that sympathin and adrenaline are inactivated 'in vivo' by oxidation to quinones, I had searched for anti-oxidants that might be of use in mammalian pharmacology, and found in 1936 that the negative oxidation

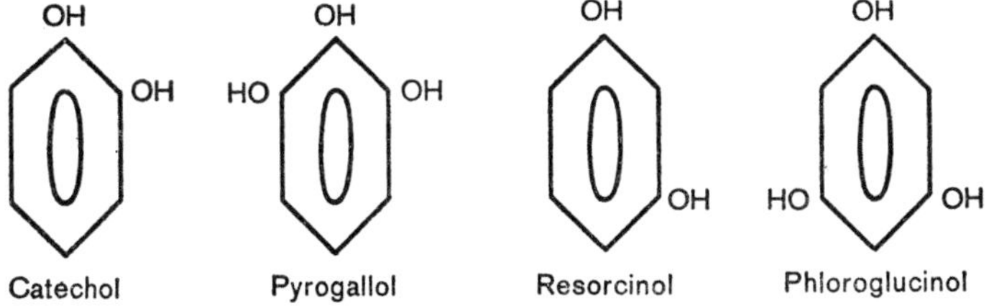

catalysts of Moureux and Dufraisse (such as catechol and pyrogallol) potentiated both the α and β effects of adrenaline and sympathetic excitation, while resorcinol and phloroglucinol (non-antioxygens) were completely ineffective. Only those polyphenols with at least two hydroxyls in the ortho position were effective. Moreover, the sensitization produced by anti-oxidants was additive with that of cocaine.

I was firmly attached to the idea of an oxidation of phenolic functions, and up until 1947 * I fought the theory held by the Oxford School (Burn, Blaschko and co-workers) which postulated that monoamine oxidase was the enzyme responsible for the inactivation of adrenaline. As so often happens, we were right on some points and wrong on others. It was the discovery of catechol-O-methyl transferase (COMT) by Axelrod and his associates which filled the vacuum. In 1949, with my colleagues, I had the opportunity to reinterpret the results of our experiments on the antioxygens, by showing that catechol and pyrogallol inhibited COMT *in vitro*. It suffices to look at the structural formulae to understand that this was due to competitive substrate inhibition, a finding which was rapidly confirmed *in vivo* by Axelrod and Laroche. Between 1937 and 1939, I began to study the changes in the effects of a dilute solution ($10^{-5} - 10^{-6}$ g/ml.) of adrenaline brought into contact with the tyrosinase of mushroom, potato or meal-worm. My secret hope was to see that the inhibitory effects (β) disappeared before the excitatory effects (α), in order to give a metabolic interpretation to Cannon and Rosenblueth's observations. For a year my co-worker Herman and I observed a totally different phenomenon. All the effects of the adrenaline solution disappeared and were replaced by a hypotensive inhibitory action. We called this derivative *adrenoxine*, although we were unable to determine its chemical structure with the techniques then available, and larger amounts of material could not be obtained since the phenomenon did not occur if more concentrated solutions of adrenaline were used. We made the most of these interesting results, which I described to Sir Henry Dale, who encouraged me, saying 'It is too good not to be true'.

Alas, I failed to demonstrate this phenomenon to the Physiological Society. Despite all my efforts to change the experimental conditions I subsequently found it impossible to reproduce the first experiments which had given me so much hope. How can

---

* See my review 'The Metabolism of Adrenaline' which was the very first to be published in Pharmacological Reviews, Vol. I, p. 1, 1947.

this terrible error be explained? Is there nothing which can lend support to the hypothesis of a metabolite of adrenaline endowed with inhibitory activity? Very little, as I indicated in my review of 1947. Some of these experiments might be explained by the presence of prostaglandins. I did not have pure enzymes at my disposal, and at that time the phenolases were not well understood. Perhaps one day someone will provide a plausible explanation.

In 1939, just before the war, I enjoyed a happy and fruitful trip to Naples with H. Fredericq. In August we were drafted into the scientific service of the Belgian Armed Forces. One could hardly give a thought to the metabolism of adrenaline. It was necessary to think about protection against chemical warfare.

9

Comparative physiology

The atmosphere in Liège when I arrived in 1932 was much oriented towards comparative physiology. Léon Fredericq told stories about his experiences in marine biology laboratories. I realized the importance of his work on Invertebrates, and Henri Fredericq, my new boss, encouraged me. Marcel Florkin had vigorously embarked on work in a direction which was to lead him far towards the understanding of biochemical evolution. In 1933 I had the pleasure to meet at Woods Hole, G. H. Parker who had already at that time extended the ideas of W. B. Cannon to the chromatophores of fish.

It was clear to me that the study of chemical transmission in the Invertebrates (that is to say the application to this immense class of creatures of the knowledge, both theoretical and practical, acquired from the Vertebrates) would provide the evolutionary theory with arguments as important and as fundamental as those derived from studies of the chemistry of haem pigments, nitrogen metabolism, or any morphological characteristics. The field was practically unexplored. Some preliminary findings indicated that the Annelids closely resemble the Vertebrates. In 1918, Fühner had described the sensitivity of the dorsal leech muscle to acetylcholine and its potentiation by eserine. Translated into the language of 1935, this fact suggested the probability of cholinergic transmission in the leech. In addition Gaskell had described the occurrence in some Annelid ganglia of certain cells that reacted to histochemical agents exactly like the chromaffin cells of the adrenal medulla. The presence of adrenaline was beyond doubt and adrenergic transmission was conceivable.

Without hesitation, I attacked the problem of cholinergic transmission in Invertebrates with the following basic premises in mind. For transmission to be *possible*, it is necessary:

1. That a pre-formed store of acetylcholine should be present in the organs since the action of a nerve impulse would be merely to release a certain amount of this substance on its arrival at the periphery.

2. That an enzyme, a cholinesterase, should be present, which inactivates the released acetylcholine and ensures that the effects of nerve impulses remain localized.

3. That the organ responds to low concentrations of acetylcholine.

To *prove* the transmission, it was further necessary to show:

(*a*) That on nerve stimulation, a substance which has all the properties of acetylcholine diffuses, in the presence of eserine, into the fluid which perfuses the organ.

(*b*) That eserine potentiates the effect of nerve stimulation.

(*c*) That one or other acetylcholine antagonist (atropine or curare) is effective, although the absence of this criterion does not necessarily exclude cholinergic transmission.

In the course of four years which included two visits to Naples and one to Plymouth, as well as some work with G. Coppée in Liège, I was able to establish the following facts which were presented in August 1937 at a discussion meeting of The Royal Society of London and in 1938 at the International Congress of Physiology in Zürich.

1. The Annelids (*Hirudo, Lumbricus, Arenicola, Aphrodita*) and *Sipunculus* meet all these conditions. Their physiology and pharmacology in the cholinergic field is the same as that of the

Vertebrates. These findings confirm the views of the classical zoologists who maintain that the Annelids and Vertebrates are closely related.

2. The Holothurians, taken as an example of the Echinoderms because of the technical ease with which we can experiment on their musculature also have cholinergic nerves. I recall that my zoology teacher Auguste Lameere considered the Echinoderms to be closely related to the Annelids.

3. Acetylcholine was extracted from Molluscan tissue. It is present in such large amounts in the central nervous system of the Cephalopods that Mazza and I, in 1935, succeeded in isolating a little more than 6 mg of acetylcholine chloroaurate from a trichloracetic acid extract of the central nervous system of 15 large Octopi (*Octopus vulgaris*).* Mollusc tissue extracts inactivate acetylcholine. But eserine does not augment the effect of motor nerve excitation on the muscles of the Gastropods, Lamellibranchs or Cephalopods. The isolated posterior salivary gland of the Octopus, perfused with sea water, releases increased amounts of acetylcholine during nerve excitation (Bacq and Ghiretti, 1952). J. H. Welsh and others have demonstrated that the innervation of the heart in the Lamellibranch, *Venus mercenaria*, has all the pharmacological and physiological characteristics of the Vertebrate cardiac vagus and is undoubtedly cholinergic. The physiology of the shell muscle of the mantle of the Molluscs is not yet well understood, and the same is true of the giant ganglionic synapses of *Aplysia*.

4. I had concluded from my results that Ascidia could not have a cholinergic innervation even though the zoologists consider

---

* The only methods available at the time were those of classical organic chemistry. No chromatography, no ultra-violet or infra-red spectra. Mazza, however, was an excellent and patient chemist. It was the first time that pure acetylcholine was chemically isolated from an Invertebrate, and it is still the only example of its isolation from an extract of nervous tissue. Recently, Whittaker isolated the vesicles which contain acetylcholine by fractional centrifugation of Cephalopod brain extracts.

their larvae close to the Amphibian tadpoles in that they possess a dorsal cord. A recent re-evaluation of the observations made on this important phylum, speaks for a chemical transmission.

5. The *Actinia*, taken as a model for the Coelenterates, have neither acetylcholine nor cholinesterase. In 1936, in Plymouth, I verified this fact with Nachmansohn, and together we published a short paper which has since been forgotten by him as it contradicted his theory which postulates that acetylcholine and cholinesterase play an important role in the propagation of the impulse along the nerve fibre, in contrast to its role at the ending, as postulated by Dale and his group. *Actinia* muscle does not contract when placed in an acetylcholine 'syrup', but does react to 5-hydroxytryptamine. It may be that 5-HT, with certain amino acids, was the first amine to be selected in the course of biochemical evolution to fulfil the role of a chemical mediator for impulse transmission across the synapse.

6. In the *Arthropods* (Crustaceans, Insects), it was soon clear that the striated muscles are not innervated by cholinergic nerves. In the Crustaceans there are two types of nerve fibres, some are excitatory and others inhibitory. It was, therefore, necessary to find two transmitters or two electronic mechanisms. What I had overlooked, however, was the importance of cholinergic transmission in the nervous system of Insects and Crustaceans. This is an unforgiveable oversight, considering the amazing sensitivity (discovered long before the 1940 war) of insects to organic phosphate-esters. However, it was not known that these substances were powerful cholinesterase inhibitors until after the war, during which the British and Americans, notably Nachmansohn and his associates, had extensively studied these substances which were manufactured as chemical warfare agents by the Nazis, and are now widely used as insecticides. At the time I was preoccupied with other problems: radiobiology and chemical protection against ionizing radiation. This took up most of my time, and to my great regret I had to leave the field after preparing a paper on adrenaline metabolism for the newly created journal *Pharmacological Reviews*. If I had looked for acetylcholine in the Crustacean nerv-

ous system it would surely have changed the direction of my work.

The present state of comparative physiology with regard to chemical transmission in the Invertebrates is summarized by E. Florey in the textbook *Fundamentals of Biochemical Pharmacology* (Pergamon Press, 1971), and in a recent and fine article by H. Fischer.* The most complete work is a volume of 1074 pages by H. Fischer entitled *Vergleichende Pharmakologie von Überträgersubstanzen in tiersystematischer Darstellung.*** My friend M. Michelson *** of Leningrad edited a section on comparative physiology (1973), for the *International Encyclopedia of Pharmacology*, in which the reader will find an excellent resumé of our knowledge in this field, which is so important for an understanding of the mechanisms of evolution. In the introduction to this volume, M. Florkin and I point out that choline esters are rather common molecules widely distributed in bacteria and certain plants; mushrooms synthesize them; choline is a molecule easily esterified by the simplest organic acids.

In the Invertebrates several choline esters have been found which in the species in which they occur are apparently more effective than acetylcholine in synaptic transmission. For instance,

---

* Animal evolution in the field of synaptic substances. *Naturwissenschaften*, 59, 425-435 (1972).

** Handbuch der exper. Pharmakol. Band XXVI, Springer-Verlag, Berlin-Heidelberg-New York, 1971.

*** I met M. Michelson in 1935 at the Moscow railway station, on the platform where most of the Western physiologists were arriving for an International Congress, at the invitation of Pavlov. He had heard of my work and wished to discuss it with me. In a few days, we became the best of friends; we see each other only rarely now, but always with the greatest of pleasure.

The story of that interminable train ride that brought hundreds of physiologists from Warsaw to Moscow should be told. Szent-Györgi went from one end of the train to the other, explaining his views on the role of tricarboxylic acids in carbohydrate metabolism, and thinking of all the young scientists whose careers would advance rapidly if the train were derailed.

urocanylcholine (or murexine) in *Murex* (Mollusc), allylcholine in *Buccinum* (gastropod mollusc) and senecioylcholine or β, β'-dimethylallylcholine in an insect (*Arctia caja*). The function of these esters in the neuromuscular systems only became possible with the progressive development of the terminal structures of the nerve fibres and of the specific receptors where the concentration of acetylcholinesterase occurs. Michelson realized very early that our interest in comparative physiology has to concentrate on the study of cholinergic receptors. The synthesis, storage, release and inactivation of acetylcholine differ very little in all animals having a cholinergic innervation. It is the fine structure of the receptors and their rapid reactions whose study requires the most elaborate biophysical techniques, and in the final analysis the receptors strategically distributed on the cell membrane determine the characteristics and modalities of the response of a cell to nerve stimulation.

Studies of the monoamines have produced numerous and interesting contributions to comparative physiology. Catecholamines

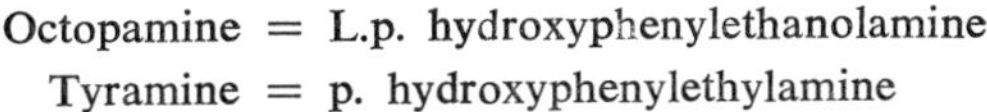

Octopamine = L.p. hydroxyphenylethanolamine
Tyramine = p. hydroxyphenylethylamine

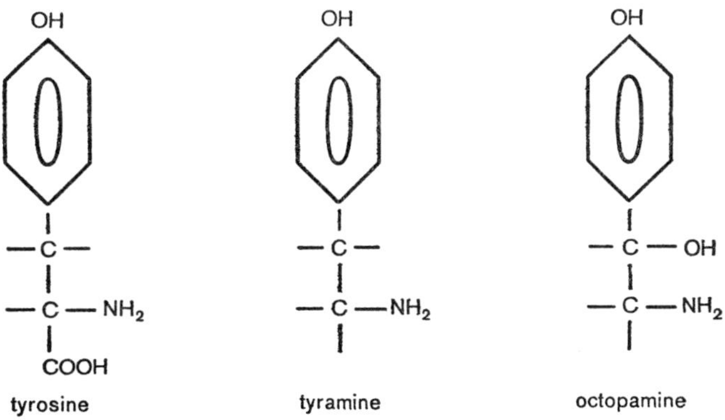

have been found in certain Protozoans and in certain cells and the tentacular apparatus of the Actinia, although their physiological significance in these cells is not clear. A large variety of monophenolic amines (tyramine, octopamine) as well as the diphenolic catecholamines identified in Vertebrates were also found in the Molluscs. Octopamine, discovered by Erspamer in 1952, is of particular interest as it shows the possibility of side chain hydroxylation of tyramine, which is not synthesized in the Vertebrates. In other words, in the *Octopus*, tyrosine is decarboxylated and hydroxylated on the side chain, while in the mammal these two operations take place *after* hydroxylation of tyrosine in position 3 of the aromatic ring to form L-DOPA.

Thanks to the wonderful work of Erspamer and his associates * it is known that 5-hydroxytryptamine is found in a number of Invertebrates, and that it probably plays an important role in synaptic transmission. Monoamine oxidase is found in the Cephalopods and Echinoderms.

Histamine is present in large amounts in the secretions of the posterior salivary gland of the Cephalopods, along with tyramine and other phenolic amines, and 5-hydroxytryptamine is found in *Octopus vulgaris*.** The central nervous system of Cephalopods is also exceptionally rich in histamine and certain pharmacological findings might indicate a physiological role for this amine.

The effects of γ-aminobutyric acid and certain other amino acids (glutamate, aspartate, glycine) have been carefully studied in Crustacea, and on the relatively simple ganglion cells of Molluscs; they may also be active in synaptic transmission.

---

* See V. Erspamer (editor): '5-Hydroxytryptamine and related indolalkylamines' *Handbuch exp. Pharmakol.* Vol. 19, Springer, Berlin-Heidelberg-New York, 1966.

** Together with Ghiretti, I found that the posterior salivary glands of the Octopus could be perfused without difficulty and the effects, on this isolated preparation of nervous excitation on internal or external secretions as well as on blood flow are easily observed (*Arch. Internat. Physiol. 59*, 288, 1951).

10

The triumph:
The central nervous system

After 1936, following the award of the Nobel Prize to Loewi and Dale, the stage was set and the theory of chemical transmission went from triumph to triumph, slowly passing into the classical textbooks of physiology and pharmacology. As we have seen, only one strubborn and immensely gifted opponent, J. C. Eccles, was left to carry on the battle. There were from time to time also some sceptics who published results that did not fit readily into the frame of the general theory and many eminent neurologists showed a sort of indifference to the quarrel between the schools of Dale and Eccles because the nature of their work was such that their experiments could be interpreted and carried out without regard to any theory of synaptic transmission.

After the autonomic nervous system, striated muscle and mammalian ganglion cells, the vast world of the Invertebrates was entered, and finally, the most difficult but also the most fascinating of all, the central nervous system. Research on chemical transmission in the CNS, which began timidly during the war of 1939-45, is still far from completion.

The technical difficulties of work in the CNS are enormous. Before 1939, it had been easy to obtain irrefutable evidence in peripheral nervous systems. Whether it was a gland or a smooth or striated muscle, the organ could be isolated by ligatures, its nervous apparatus precisely stimulated and the venous blood or perfusate could be collected and analyzed. Transmitter substances could be injected into the artery. The situation is very different in the CNS: a mass of neurones and fibres, packed with glial cells, whose circulation is impossible to control and containing 'centres' which receive many different inputs and which are more or less arbitrarily localized by anatomists and physiologists.

With the methods in common use, which in the meantime had become nearly classical, it was shown that acetylcholine, catecholamines and the cholinesterases were present in the CNS. This painstaking work has allowed us to draw up a kind of chemical map. Acetylcholine is found almost everywhere, but most concentrated in the cerebral cortex; a highly specific and sensitive test for acetylcholinesterase * reveals that this enzyme is concentrated at certain synapses, but is found in only minute quantities at others. Noradrenaline is found in a number of centres but dopamine, the immediate precursor of noradrenaline has been found in substantial amounts only in a few well-defined areas. 5-Hydroxytryptamine has also been identified in various centres.

The neurone theory established by Cajal, the famous Spanish histologist, is confirmed biochemically. The neurone, its axon and endings, synthesize the transmitter which is stored in specialized vesicles which are visible with the electron miscroscope. These vesicles burst under the influence of the nervous impulse arriving at the nerve terminal, releasing their content into the synaptic cleft. The granules are formed in the cell body of the neurone, and filled with acetylcholine or amine molecules, are transported along the axon at a speed which in certain situations can be experimentally determined.

---

* The principle of this reaction is to incubate layers of frozen tissue with acetyl- or butyrylthiocholine in the presence of a copper or lead salt. The insoluble mercaptide produced by the liberation of thiocholine is transformed into copper or lead sulphide which does not diffuse and is easily visible in the light microscope. With certain modifications of the technique, the electron microscope can also be used.

New techniques are establishing the experimental evidence needed to confirm the nature of the transmitters. Feldberg and many other experimenters have succeeded in implanting in the cat and rabbit brain a permanent cannula, the tip of which lies in the third or fourth ventricle or in one of the lateral ventricles. Thus substances can be brought in close contact with the nerve cells which line these cavities and their effects on the non-anaesthetized animal can be observed. In this way, the blood-brain barrier is circumvented. Such cannulae also permit the removal of cerebrospinal fluid for microchemical examinations. Microelectrodes have been perfected for the recording of potentials either at the membrane surface or intracellularly from nerve cells. By using a stereotaxic apparatus one is able to place the tip of an electrode in any desired place and the functional role of the pierced neurone can be determined by various types of peripheral stimulation. The injection of a small amount of dye at the end of the experiment enables the anatomist to confirm the localization of the electrode, using either light or electron microscopy.

Fine glass tubes filled with the substance whose action is to be studied can also be inserted with electrodes; if the substance is an electrolyte one can determine the exact quantity liberated by electrophoresis during application of a direct current of short

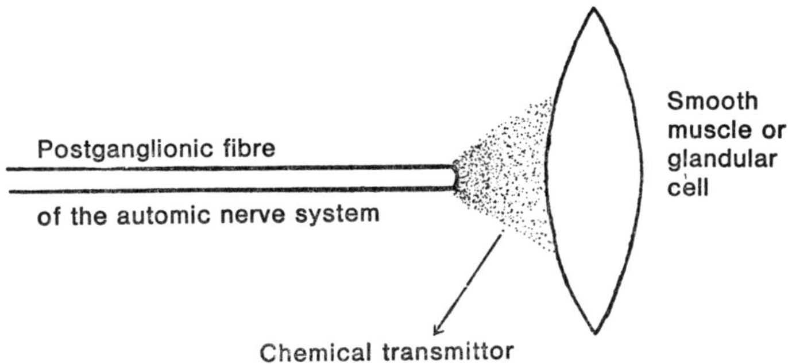

FIG. 13 *a*. — Schematic presentation proposed by Bacq in 1934.

duration, which will simulate the discharge of transmitter from the presynaptic apparatus. The responses are recorded and compared to potentials evoked by physiological stimuli; with several tubes in multibarrel microelectrodes consecutive injections of minute quantities of different substances can be introduced; this

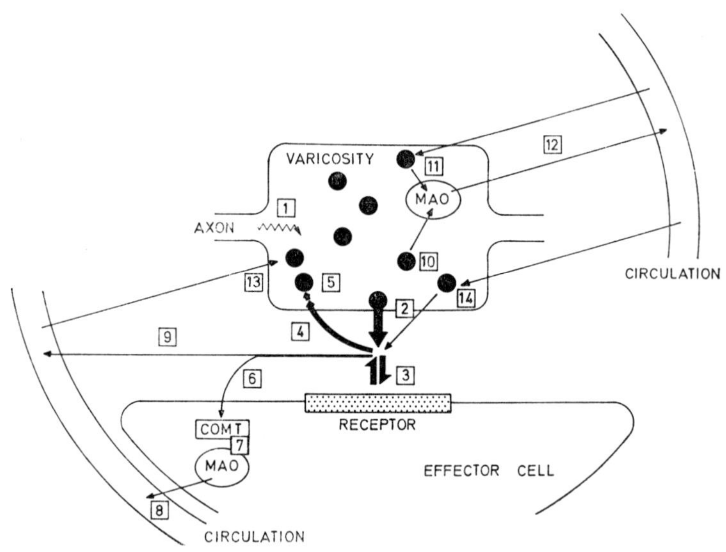

Fig. 13 *b*. — Sites of action of drugs at an adrenergic synapse. (1) Propagation of action potential in presynaptic terminal; (2) release of noradrenaline (NA); (3) interaction of NA with α- or β-receptors; (4) neuronal uptake of released NA; (5) uptake of axoplasmic NA into storage vesicles; (6) uptake of released NA into extraneuronal cells; (7) catabolism of NA in extraneuronal cells; (8) loss of NA catabolites to circulation; (9) overflow of unchanged NA to circulation; (10) spontaneous loss of stored NA to intraneuronal monoamineoxidase (MAO); (11) reserpine-induced loss of stored NA; (12) loss of deaminated catabolites to circulation; (13) uptake of circulating catecholamines into adrenergic terminal; (14) displacement of stored NA by exogenous sympathomimetic amines.

Iversen and Callingham, in *Fundamentals of Biochemical Pharmacology*, Bacq ed., Pergamon Press, 1971, p. 304.

makes it possible to analyze the effects of the classical agonists or antagonists. An incredible wealth of information is now available to neurophysiologists and neuropharmacologists. Imaginative organic chemists, stimulated by the specialized pharmacologists are continuously providing the experimenter and the clinician with interesting new compounds of one sort or another. Figure 13 gives an idea of the progress made during the last forty years.

In the feverish joy of research, pharmacologists and clinicians have had a tendency to forget that the basis of their work had already been established in 1940. The fertility of the theory of chemical transmission is not yet exhausted. Certain amino acids, besides acetylcholine, noradrenaline, dopamine and 5-hydroxy-tryptamine are also considered likely to be excitatory transmitters, and γ-aminobutyric acid (GABA) plays the role of an inhibitory transmitter of some synapses both in Vertebrates and Invertebrates.

11

Working in
Sir Henry Dale's department

For a physiologist already well trained as I was in 1936, it was an unmitigated joy both physical and spiritual to spend some time in an extraordinarily well organized laboratory. Sir Henry, after a brief visit to Liège, enabled me to obtain a grant from the Rockefeller Foundation, and I spent a most happy and fruitful six months in London and Plymouth.

Dale's principal associate at the time was G. L. (later Sir Lindor) Brown.* W. Feldberg was in Australia at the time, at the Walter Eliza Hall Institute, whose Director was C. Kellaway. G. L. Brown and I rapidly became close friends. I worked with him practically every day at the National Institute for Medical Research. He was competent, scrupulous, generous, the enemy of pomposity and conscientious in all aspects of life and all functions he assumed, notably as Secretary of the Royal Society. The worst criticism he had of a colleague who was too preoccupied with publicity was: 'He is too full of himself'. One day at University College I took part in a tutorial which he gave to a small group of students. What kindness! No question of a contest, everyone said what he thought. His manual skill was extraordinary, he had no equal

---

* An exceedingly good biography of George Lindor Brown has recently been prepared by two of his closest friends, F. C. MacIntosh and W. D. M. Paton and published in the *Biographical Memoirs of Fellows of The Royal Society*, Vol. 20, 1974, 41-73.

in inserting a cannula into a minute vessel, or in mastering the most minute technical details.

The success of extremely difficult experiments was assured from the outset. The experiment might give a negative result but it rarely failed. What one has to do in experimental physiology is to put questions to the animal; no clear reply can be expected if the question is not simple, precise, and logical. Each step is discussed; one is not discouraged by difficulties. Each experiment is so well done that it rarely needs to be repeated; the tracing obtained is always so good that it can be used for publication. Not that in W. B. Cannon's laboratory the techniques were not also impeccable. No amateurish work! Brown and I often took an hour and a half, or even two, to carefully prepare an animal; after that the experiment itself went quickly, it was not unusual for everything to be finished within half an hour.*

Statistics had no place in this method, which allowed rapid progress at a regular rhythm most pleasing to the mind, a method of 'relaxation' as one would say today, which left time for thinking and discussion. Many young 'Continental' physiologists have told me that they were surprised by the fact that their British colleagues gave the impression of working very little, especially considering the number and value of their publications. There is no secret other than the quality of the work!

Dale's method of working became clear to me shortly after my arrival in his laboratory. I told him of the problem of veratrine.** Just as with eserine, veratrine also potentiated the effect of motor nerve stimulation on skeletal muscle, but I had shown that veratrine had no cholinesterase inhibiting activity. When I

---

* We also owe him and his associates J. S. Gillespie, C. B. Ferry and A. G. H. Blakeley a method for perfusion of the isolated cat spleen with the heparinized blood of the cat, a method which has proved most useful for the study of noradrenaline release by adrenergic nerve stimulation and also the mechanisms of uptake of catecholamines.

** At that time the alkaloids had not yet been separated, veratrine being the naturally occurring mixture.

demonstrated this fact to him in a simple experiment Dale reacted to it with incredible swiftness. I was under the impression of watching an almost instant reflex reaction of an outstanding mind, of a brain aware of all the essentials which had to be considered and proved. This mind, which worked with faultness logic, better than any modern computer, provided the starting premises and indicated the experiments that had to be done, and their probable results; the paper could have been written before starting the experiments. One merely needed good tracings for publication. The experimental design was decided upon in a few minutes by Sir Henry, G. L. Brown and myself and was executed within a fortnight: the action of veratrine had nothing to do with that of eserine, it was an action on the muscle fibre, it persisted on striated muscle stimulated directly, nine days after denervation, even though at this time eserine potentiation was abolished; ether anaesthesia and curare which suppress the action of eserine did not abolish that of veratrine; the action potentials of veratrine-sensitized muscle are abnormal and asynchronous (Bacq and Brown, 1937). I showed later, in 1939, with Szent-Györgyi and Goffart that the mode of action of veratrine is essentially ionic.

The quality of a physiologist's work is assured first of all by focussing on a precise technique after discussing every detail. In fact, generally after several weeks the various experimental procedures are carried out more routinely, which is very satisfactory, as repetition increases precision. However, the experimenter must be able to justify, at each stage, each step in his design; the choice of animal, anaesthesia or the use of the spinal animal, the entire set-up for recording, the frequency of stimulation, the dose of a drug, etc. Brown and I in 1937 understood very clearly why Rosenblueth and Morison had failed in 1936 to observe the important phenomenon of eserine potentiation of the response of striated muscle to nerve stimulation which Brown, Dale and Feldberg had already observed in 1935. They had used only one frequency of stimulation, which was too high; their anaesthesia and the use of too high a dose of eserine were also unfavourable.

The elegance of a technique often requires a good deal of manual dexterity. When one comes to a new laboratory, no-one will ask you 'Are you good with your hands?' They just observe you while you are working. The clumsy ones are soon found out; there is a tendency to believe that they are not bright and that they will not go far. For us, science is a craft which has much in common with art. A piece of scientific research has the same uniqueness as a work of art. When one speaks of a 'beautiful experiment' it is up to a certain point, really the aesthetic beauty that is implied. In experimental science as in art, freedom of thought is possible only after the technical difficulties have been overcome. Those who imagine magnificent experiments which they cannot undertake because of a lack of technical means or of manual dexterity end up by suppressing their imagination and forgetting their ideas, a sad way of avoiding depression. For some the ideal way is to do the experiment on the very day that they thought of it, while jumping out of bed. Of course, some fundamental phenomena require months or even years for their clarification. Patience is the virtue required of those who do this kind of work. Ill luck sometimes overtakes the experimenter who realizes at the end when the results are analyzed that an important control is missing. Physiology has its sprinters, its middle distance and its marathon runners.

12

Scientists and science

The development of the theory of chemical transmission abounds in lessons for the history of science. It shows how an idea, revolutionary at its inception, progresses slowly, thanks to the work of a few exceptional scientists who devoted their time and energy during a long period of their life to its elucidation, and put all the weight of their authority in the balance. These men, unevenly distributed in the technically developed countries at the time covered by this story (1920-50), had certain distinguishing marks in common. All were satisfied with their work, serene and without complexes. But their temperaments were different, and one could see the expression of this in their working methods as well as in their style of life and in their writings.

12.1. EUROPE VERSUS THE U.S.A.

Most of the experimental work and exciting discussions on the theory of chemical transmission took place in Europe. W. B. Cannon and his group at Harvard are the only ones in the United States who made important contributions. J. C. Eccles (now Sir John) based in Australia or New Zealand brought opposing views to the meetings of the Physiological Society in Great Britain. In the British *Journal of Physiology* the number of papers between 1926 and 1941 dealing with acetylcholine was as high as 115. In the *American Journal of Physiology* during the same time period, it was 63; and between 1926 and 1929 one finds only one insignificant publication on acetylcholine.

What strikes the impartial historian is (1) the immense contribution of British physiologists and pharmacologists, helped by German exiles after the first impetus had been given by an Austrian scientist; (2) the important contribution of the Swedish, the more modest one of the French; and (3) the recent overwhelming influence, but in a different direction, of the American biochemical pharmacologists. There is no field of science where British tradition has given a better demonstration of its virtues; the present generation of the younger great leaders in physiology and pharmacology is as remarkable as the one which, from 1900 to 1950, built the reputation of British science.

Twenty-two years ago (September 10 and 11, 1953) a symposium on chemical transmission was held in Philadelphia.* O. Loewi, Sir Henry Dale, sixteen British (or naturalized Germans), six Scandinavians and four other Europeans presented a paper; Koelle was the only one to represent, very well indeed, the U.S.A. R. W. Gerard from Chicago, who had accepted the difficult task of giving the closing remarks, began his speech with the following sentences: 'It is not entirely clear to me why I was asked to summarize and close this programme. As the lone American in this distinguished galaxy and as one of the few physiologists amidst the pharmacological cohorts, the Committee may have wished me to serve as devil's advocate for an electrophysiological approach to neural functioning . . .' He spoke at length in favour of 'electrical' transmission using many arguments the weakness of which is now obvious, but which at the time still carried some conviction.

In October 1958, just five years later, the geographical distribution of participants in the symposium on catecholamines had completely changed. At Bethesda (Maryland) at the First International Catecholamine Symposium, organized by the N.I.H. the number of contributions (invited reviews, individual papers and discussions) from U.S. scientists was twenty-nine compared to four British, three Swedish, one Canadian, and one German.

---

* *Pharmacological Reviews*, 1954, *6*, 1-131.

During the years 1960-70, the Swedes developed a vigorous counter-offensive and, alone with the British, challenged the American supremacy. In the Third International Catecholamine Symposium at Strasbourg (June 1973) slightly more than half of the papers were by scientists from the United States; the Swedes, again, were second.

12.2. THE ADVENTUROUS SPIRIT; THE FLAIR

In spite of their peaceful and conventional appearance (which in the language of young Marxists would be called 'petit bourgeois') the scientists whose work is discussed here are true adventurers. They considered themselves as such. The dedication which Sir Henry wrote in 1954 on my copy of his 'Adventures in Physiology' which I have frequently quoted, reads as follows: 'To my friend and, for a time, my fellow adventurer, Z. M. Bacq with all good wishes for the success of his further expeditions'. But if the scientist is an adventurer, he also shows prudence and caution. He knows what his colleagues think. He is acutely aware of what happens in his field all over the world. If he expresses revolutionary ideas without hesitation, he may sometimes also retreat, or at least stop plunging headlong in the direction he himself first proposed. For instance, T. R. Elliott, in his magnificent paper on the action of adrenaline (*J. Physiol.* 1905, *32*, 401) does not express himself with the same clarity and extraordinary lucidity that one finds in his note of 1904 to the Physiological Society. Dale aslo acknowledges that in his paper on the action of ergot (1906), he did not mention Elliott's hypothesis, although at that time Elliott was already a dear friend of his and had discussed the content of this note with him many times. Dale admits (see *Adventures in Physiology* p. 35) that he was not 'ideally courageous'.

The inner struggle of the scientist exists not only between the adventurer and the man of caution, but also between his inquisitiveness and his laziness. One of my best friends, a good honest scientist, considers himself to be genetically lazy and born to live like the pashas of oriental legends. But he is prey to such implacable and relentless curiosity that his existence has been dominated by one publication after another, and he gives his colleagues the impression of being one of the most active men of his time. One of the major qualities of the scientist, and particularly of the biologist, is to grasp the significance of unexpected facts which experiments put by chance into his hands and not allow them to escape. The reader can find on page 38 how a number of chance observations and their logical interpretation led W. B. Cannon to perform the decisive experiments on 'sympathin' with me. Dale, whose flair was rarely wrong, tells us of two astonishing incidents of what is called 'serendipity'. It was by chance, a chance which luckily happened again a week later, that Dale first observed the reversal of the pressor effect of adrenaline after injection of an extract of ergot. It was also by chance * that Dale found in a putrified extract of ergot the three substances which became the subject of fifty years of research (see Dale, *Adventures in Physiology*, p. XIII and following pages). First the sympathomimetic amines (in particular tyramine) which Barger and Walpole had already found in putrified meat; second, histamine which, like tyramine had been produced by bacterial decarboxylation of the corresponding amino-acid, histidine in the case of histamine; thirdly, acetylcholine which A. J. Ewins had succeeded in 1913 in isolating with the very simple methods available at that time. Fresh ergot does not contain any acetylcholine. It is probably thanks to a contamination by *Bacillus acetylcholini* that the accidental presence of acetylcholine was observed in these extracts. This micro-organism is found in sauerkraut and in fermentation silos; it has the necessary enzymatic equipment to synthesize acetylcholine. As it has no cholinesterase, the acetylcholine formed is relatively stable and diffuses freely into the medium. Dale's merit was that

---

* The chance here is that at some stage in the preparation of the ergot extracts made by the Wellcome Company for medical use they must have become contaminated by bacteria. Sir Henry worked at the Wellcome Physiological Laboratories from 1904 to 1914. In those days it was something unheard of, as Dale himself said, for a scientist to take up a position in a pharmaceutical firm.

he differentiated between the actions of histamine and those of acetylcholine and realized that the presence of the amines derived from decarboxylation of the amino acids could not account for all the biological effects of this putrified extract of ergot.

12.3. The power and simplicity of facts and their utilization

Facts are omnipotent. The development of a precise technique the results of which can be easily interpreted takes time, and requires much patience, skill and imagination. Once done, it satisfies the mind as well as the hands, like a concerto written for the virtuoso. The interpretation is seldom doubtful. To be able to discuss is one of the great pleasures of working with colleagues; everyone contributes. To discuss without losing oneself in flights of useless, pompous and sterile rhetoric; to state precisely the general direction in which the work should proceed and to readjust this frequently; to choose which experiment comes next; to draw conclusions from each experiment—good or bad—and to extract from it the maximum of information; to compare each day's result with the previous ones as well as with the older data already available (and here a good memory is always a great help): these were the 'intellectual' processes of the mind I witnessed and took part in. They were often brilliant and rapid but sometimes the useful reaction came only after a long latency. The most lively mind is not always the most effectual. A certain slowness is often associated with greater reliability and greater determination to carry out the experiments which thought has suggested.

It is within the most classical rationalism, the Cartesian rationalism exemplified by Claude Bernard—that Loewi, Dale and Cannon succeeded with their associates in laying the foundations of the chemical transmission theory. They were able to make the most of the resources of physiology, pharmacology and elementary biochemistry available before 1940. Many readers of this historical sketch will find it difficult to accept the fact that these foundations were established without the help of statistical methods. W. Feldberg confirmed that in this matter his recollections coincide with mine. Not that Dale and his coworkers were opponents of these methods, but the truth is, they found them useless for the kind of research they were pursuing. Among Dale's associates, J. H. Gaddum often resorted to mathematical methods. Dale himself was very active and successful in devising standards for hormones and vitamins at a time when these substances were still impure or chemically unidentified. The experiments of Loewi, Dale or Cannon were conceived in such a way that the result was simple and clear, positive or negative. One should use statistical methods only if one is forced to do so, then one must choose carefully the best one. Prolonged and often sterile discussions surround facts which are considered by some to be statistically significant, while others maintain that they are meaningless. The following story is attributed to Jean Brachet: sitting at dinner near a well known American statistician, he deliberately showed his lack of interest for statistical methods. 'But' said the expert, 'when you get results like these'—he grasped the invitation card in front of him and wrote hastily a few disjointed data—'how can you avoid using statistical methods?' Brachet looked at the card and then came the answer: 'If I get results like these, I would just throw them in the wastepaper basket and start again.'

When a short paper was sent back to me by an American journal because I was chided for not having done the statistical analysis of a series of experiments in which 97 % of the treated animals survived whereas 98 % of the controls died, I merely added to the text, 'No statistical analysis is needed to show that the results are significant.' The paper was accepted and published. Perhaps the editor failed to realise that this addition was an insult to the intelligence of the reader?

Statistical analysis does not improve faulty observations. Some people believe that they do good work if they multiply experiments and let them be carried out by their technicians without supervision, instead of perfecting their methods and performing the experiments themselves. For instance, a carefully performed unilateral sympathectomy allows the comparison, in the same animal of the reactions of the normal and the denervated side. At the present time, in preference to this long and delicate technique, a large number of rats are often injected with various drugs (reserpine, 6-OH dopamine) which induce a kind of 'chemical sympathectomy'. Naturally the variability of the results obtained is such that statistical analysis becomes necessary. Far be it for me to deny the usefulness of these substances, which penetrate into the brain where the facilities offered by experimental surgery are limited. Each drug is a precious tool for research as soon as its biochemical mode of action has been ascertained. But why give up the more elegant and more appropriate classical methods? Only because they are more difficult?

Some biophysicists and molecular biologists, as well as some theorists would nowadays like to make something grandiose out of biology, an impressive speculative science, capable of arousing and retaining the enthusiasm of the crowd. I do not believe this is possible or even desirable. The enthusiasm of the crowd is more easily and more frequently roused by more colourful stories; astrology will always gather more supporters than biochemical pharmacology. Furthermore, the level of taste to which many journalists expose the public is deplorable. Perhaps one cannot blame them. How can they convey to their readers what science involves—that it takes hours day by day over many years in order to appreciate the concepts, the amazing beauty and serenity which emanates from contemporary science.

Philosophers and writers often neglect the possibilities open to them by unfashionable theories. To my knowledge neither Marxists, supporters of Teilhard de Chardin or Popper, nor the theoreticians of molecular biology have systematically exploited the abundance of facts brought to light by the theory of chemical transmission. The reason may be that this field lies a little away from the highways on which they travel with their racing cars. Yet what an admirable conquest! What beauty in this varied utilization by living organisms of minute amounts of a few simple molecules!

12.4. The diversity of temperaments

One would not easily find more varied characters and more different personalities than Otto Loewi, Sir Henry Dale, W. B. Cannon and Sir John Gaddum. I have already frequently referred to Sir Henry and his group; I shall complete his psychological profile by adding that until his death at the age of ninety-three, Sir Henry retained all his faculties and preserved a remarkable sense of humour.

Otto Loewi had studied medicine in Strasbourg before the 1914 war. At the beginning of the century he was professor of pharmacology at the University of Graz (Austria). Because of his open and firm character he did not conceal either his Jewish extraction or his new and intelligent views about the position of pharmacology in the basic medical sciences. Nowadays we go further than him and realise that pharmacology is one of the great disciplines in biology. Fortunately, Loewi left an autobiographical sketch in which humour blends with vigour and power. In his time, papers published in German were read by all concerned and had the greatest influence. Lembeck and Giere have published a book on Loewi with biographical documents and a critical analysis of his scientific work which is so complete and so admirably presented that it is impossible to add anything but a few personal recollections.

I saw Loewi for the first time in Belgium in 1939, a few months before the war broke out. He had arrived alone, without any money, a real refugee. The Nazis had allowed him to leave Austria only after he had left the total amount of his Nobel Prize in their hands. The Fondation Francqui had offered him a professorship with us and he was given an assistant, Dr Dulière. He made highly appreciated visits to the various Belgian universities, and had discussions with many young scientists, in this way contributing to their 're-cycling'.

I remember his visit to Liège where I demonstrated for him the hypersensitivity to potassium ions induced by veratrine in toad muscle. Two comments from him: 'I would love to have discovered that. You must study this phenomenon carefully, quantitatively, the main thing, quantitatively.' At a dinner given by the Foundation in his honour, he pleasantly commented upon the Freudian interpretation by Dulière of his famous primordial experiment: 'to bring something, a subtle matter from one heart to another...'

I met Loewi again in 1957 in New York with his wife who had managed to join him there in 1941. He invited my wife and myself to dinner in a sumptuous restaurant. His health was precarious. Mrs Loewi looked anxious after his fits of breathlessness which interrupted the conversation, but he was extremely happy to meet an old comrade in arms of pre-war years. 'Bacq' he said 'I should have been dead long ago. But everything interests me, I do enjoy living on overtime.' * He gave the impression of indomitable energy, like a force of nature, a kind of volcano throwing out its last fires before becoming extinct.

Walter Bradford Cannon, Georges Higginson Professor of Physiology at the Harvard University School of Medicine possessed all the fine qualities of the American people; he was unostentatious and had a natural and charming modesty which endeared him to his European and Chinese friends. One found in him nothing but unselfishness, generosity, kindness, but also a great strength of mind. These qualities, coupled with an astonishingly young personality were the reason why he was surrounded, without any effort on his part, by co-workers and friends who had the deepest affection for him. Unavoidably, such a personality could not fail to excite a good deal of controversy which forced W. B. Cannon to fight vigorously for his ideas, but never without humour. Nothing was more characteristic of him than the important part he played in the years 1936-39 in

---

* His wife translated the German expressions he used—his conversation was partly in English, partly in German.

Another typical remark of Loewi to Henri Fredericq who told him that Nazi Germany had regressed to the Middle Ages—'Don't speak ill of the Middle Ages'.

supporting the cause of the Spanish republicans in the States. In 1935 at the International Congress of Physiology in Moscow and Leningrad—where he met his dear friend I. P. Pavlov for the last time—he understood that fascism and nazism lead straight to war. In the plenary session, he delivered a splendid eulogy to liberty, which greatly offended the delegates from some countries. Until the very last days of his life, the necessity of maintaining friendly contact between American and Soviet scientists was one of his main preoccupations.

Three years ago, at the initiative of Professor Chandler-McC-Brooks, a symposium was organized at New York University to commemorate the hundredth anniversary of W. B. Cannon's birth. All his surviving friends, associates and members of his family were present. The proceedings have now been published.* Eminent speakers recalled that in 1894 W. B. Cannon had the idea of administering by stomach tube a substance opaque to X-rays (a bismuth salt) in order to watch the movements of the stomach and intestine in the unanaesthetized cat; they recalled that he was the great pioneer in the study of the physiological effects produced by emotions and that his concept of homeostasis has been so useful that many physiologists and biologists today use it without knowing that it was W. B. Cannon who forged and tempered it. In my opinion, which is a very personal one, W. B. Cannon does not enjoy the reputation he deserves from his work in his own country; his work is valued more highly in Europe. A kind of silence has built up around his personality, because his life, his character and his extrascientific activities did not correspond to the accepted standards. A touch of non-conformism and the fact that one was right too early may be sufficient reasons to obliterate the memory of certain exceptional men.

---

* The Life and Contributions of Walter Bradford Cannon 1871-1945: His influence on the development of Physiology in the 20th century. Edited by Chandler McC Brooks, Kiyomi Koizumi and James O. Pinkston. State University of New York, Downstate Medical Centre, 1975.

John Henry (Sir John) Gaddum was a man of exquisite intelligence and sensitivity, an adorable and in fact an adored being, not only by his family, but also by his associates, colleagues and friends of all ages. One could not miss the atmosphere of poetry and charm which surrounded him. One could not fail to admire his devotion to scientific research and his rigorous mind, which nevertheless did not exclude imagination. His house was always open to his numerous friends and their families, and Lady Gaddum and her daughters entertained with unequalled and simple cordiality.

I met Gaddum for the first time in 1935 at the International Congress of Physiology in Moscow and Leningrad. The President was Pavlov. This was one of the first contacts scientists of the Western world had with their U.S.S.R. colleagues. I can still hear Gaddum complaining about the lack of controls in so many of the experiments made by the Russian physiologists of that time.

W. Feldberg has written an excellent biography of Sir John (*Biographical Memoirs of Fellows of The Royal Society*, vol. 13, Nov. 1967, pp. 67-77). Gaddum was taken on by Dale when he was just twenty-seven, and he took part at the National Institute for Medical Research in the work on chemical transmission without giving up the problem of biological standardization which, together with the quantitative treatment of the phenomena of *antagonism,* remained in the foreground of his interests. His teaching career began in 1934 in Cairo where he accomplished the near miracle of continuing his research with J. Barsoum and M. A. Khayyal. Back in London the next year, 1935 (University College and then the College of the Pharmaceutical Society) he succeeded A. J. Clark in 1942 to the Chair of Pharmacology at the University of Edinburgh. A. J. Clark was one of the outstanding pharmacologists of his time; his review entitled 'General Pharmacology' and his book 'Action of drugs on cells' gave pharmacology, until then considered the mere handmaid of physiology and clinical medicine, its title of nobility.

With the help of Marthe Vogt, Gaddum made his Institute in Edinburgh one of the great centres for pharmacological research. He ended his career as Director of the Institute of Animal Physiology at Babraham (near Cambridge) where he continued his work with Marthe Vogt, mainly on the central nervous system.

His textbooks and reviews are pleasant to read because they are written in a modest and humorous style. Feldberg has extracted a few delectable quotations from his writings. For instance:

'Enthusiasm is not enough.'

'It has always been easy to make clinical observations on the action of purgatives.'

'It will probably always be more important to try a thing out than to argue about it.'

'The pharmacologist has been a "jack of all trades", borrowing from physiology, biochemistry, pathology, microbiology and statistics—but he has developed one technique of his own and that is the technique of bioassay.'

Gaddum also had a gift (very characteristic of the true physiologist or pharmacologist) for inventing and building very simple pieces of apparatus and for devising easy methods which were quickly adopted; some of them are still widely used today. For instance, his technique of 'Superfusion' (1953) which consists of letting a fluid (blood, plasma, lymph or saline solution) drop slowly over an isolated smooth muscle, the sensitivity of which allows the detection of certain substances (cholinomimetics, catecholamines, polypeptides, or kinins) at very low concentration. The spectacular development of this technique by J. R. Vane and its application to the study of prostaglandins has given results which are as valuable as those obtained by the most sophisticated biochemical methods.

The memory of Sir John is carefully kept alive by the British Pharmacological Society which has established a Foundation bearing his name.

A few complementary references

BACQ Z. M., 'Les propriétés biologiques et physico-chimiques de la sympathine comparées à celles de l'adrénaline', *Arch. Intern. Physiol.*, 1933, *36*, 167, thèse d'agrégation.

BACQ Z. M., 'La pharmacologie du système nerveux autonome, et particulièrement du sympathique, d'après la théorie neurohumorale', *Annales de Physiol.*, 1934, *10*, 467.

BACQ Z. M., 'La transmission chimique des influx dans le système nerveux autonome', *Ergebn. d. Physiol.*, 1935, *37*, 82-185.

BARGER G. and DALE H. H., 'Chemical structure and sympathomimetic action of amines', *J. Physiol.*, 1910, *41*, 19.

CANNON W. B., *The way of an investigator*, W. W. Norton Cy, New York 1945.

CANNON W. B. et BACQ Z. M., 'A hormone produced by sympathetic action on smooth muscle', *Amer. J. Physiol.*, 1931, *96*, 392.

CANNON W. B. et ROSENBLUETH A., 'Sympathin E. and Sympathin I', *Amer. J. Physiol.*, 1933, *104*, 557.

CLARK A. J., *The mode of action of drugs on cells*, Arnold and Co, London 1933.

CLARK A. J., *General Pharmacology*, Handbuch der exp. Pharmak., Ergänzungswerk, IV Band, J. Springer, Berlin 1937, 228 p. This monograph has been recently reprinted without any revision, by Springer.

DALE H. H., 'Chemical transmission of the effects of nerve impulses', *Linacre Lecture*, 5 mai 1934, *Brit. Med. Journ.*, 1934, p. 1-20.

DALE H. H., *Adventures in Physiology*, with excursions in autopharmacology. A selection from the scientific publications of Sir Henry Hallett Dale, with an introduction and recent comments by the author, *Pergamon Press*, London, 1953.

De ROBERTIS E., 'Histophysiologie des synapses et neurosécrétions. Monographies de Physiologie causale', *Gauthier-Villars*, Paris, 1964.

Du Bois Reymond E., *Gesammelte Abhandlung der allgemeinen Muskel- und Nervenphysic*, 1877, *2*, 700. See also Langley J. N., 1906.

Eccles J. C., *Facing reality*, 'Philosophical adventures by a brain scientist', Heidelberg Science Library, Springer Verlag, 1970, 210 p.

Euler U. S. von, *Noradrenaline*, Ch. C. Thomas, Sprinfield, Ill., 1956.

Florkin M., *Molecular approaches to phylogeny*, Elsevier, Amsterdam, 1966.

Florkin M., *Concepts of molecular biosemiotics and of molecular evolution*, in 'Comprehensive Biochemistry', vol. 29A, Elsevier, Amsterdam, New York, 1974.

Fundamentals of Biochemical Pharmacology, Z. M. Bacq ed., Pergamon Press, 1971, 659 p.

Heymans C., 'Les substances anticholinestérasiques', *Exposés annuels de Biochimie médicale*, M. Polonovski, ed. Masson, Paris, 1951, p. 21-53.

Kahn R. H., 'Über humorale Übertragbarkeit der Herznervenwirkung', *Pflügers Arch. f. ges. Physiol.*, 1926, *214*, 482.

Langley J. N., 'On nerve endings and on special excitable substances in cells', *Croonian Lecture*, 1906, *Proc. Royal Soc. B.*, 1906, *78*, p. 183.

Lembeck F. und Giere W., *Otto Loewi*, 'Ein Lebensbild in Dokumenten', Springer-Verlag, Berlin, Heidelberg, New York, 1968.

Loewi O., 'An autobiographic sketch', *Perspectives in Biology and Medicine*, IV, University of Chicago Press, 1940. This text is reprinted in the book of Lembeck and Giere.

Michelson M. J. et Zeitmal E. V., *Acetylcholine*, 'An approach to the molecular mechanism of action', Pergamon Press, Oxford, 1973, 231 p.

Nachmansohn D., *Chemical and molecular basis of nerve activity*, Academic Press, New York, 1959, 235 p.

Parker G. H., *Humoral agents in nervous activity*, 'With special reference to chromatophores', Cambridge, University Press, 1932, 79 p.

Rosenblueth A., *The transmission of nerve impulses at neuroeffector junctions and peripheral synapses*. M.I.T. Technology Press; Wiley and Sons, New York, 1950, 325 p.

Rosenblueth A., *Mind and brain*, 'A Philosophy of Science', M.I.T., London and Cambridge (Mass.), 1970, 128 p.